Microelectronic Systems
Level II

Units in this series

Contents

Preface

This book is one of a series on microelectronics/microprocessors published by Hutchinson on behalf of the Technician Education Council. The books in the series are designed for use with units associated with Technician Education Council programmes.

In June 1978 the United Kingdom Prime Minister expressed anxiety about the effect to be expected from the introduction of micro-processors on the pattern of employment in specific industries. From this stemmed an initiative through the Department of Industry and the National Enterprise Board to encourage the use and development of microprocessor technology.

An important aspect of such a development programme was seen as being the education and training of personnel for both the research, development and manufacture of microelectronics material and equipment, and the application of these in other industries. In 1979 a project was established by the Technician Education Council for the development of technician education programme units (a unit is a specification of the objectives to be attained by a student) and associated learning packages, this project being funded by the Department of Industry and managed on their behalf by the National Computing Centre Ltd.

TEC established a committee involving industry, both as producers and users of microelectronics, and educationists. In addition widespread consultations took place. Programme units were developed for technicians and technician engineers concerned with the design, manufacture and servicing aspects incorporating micro-electronic devices. Five units were produced:

Microelectronic Systems	Level I
Microelectronic Systems	Level II
Microelectronic Systems	Level III
Microprocessor-based Systems	Level IV
Microprocessor-based Systems	Level V

Units were also produced for those technicians who required a general understanding of the range of applications of microelectronic devices and their potential:

Microprocessor Appreciation	Level III
Microprocessor Principles	Level IV

This phase was then followed by the development of the learning packages, involving three writing teams, the key people in these teams being:

Microelectronic Systems, I, II, III	— P. Cooke
Microprocessor-based Systems IV	— A. Potton
Microprocessor-based Systems V	— M. Morse
Microprocessor Appreciation III	— G. Martin
Microprocessor Principles IV	— G. Martin

The project director during the unit specification stage was N. Bonnett, assisted by R. Bertie. Mr Bonnett continued as consultant during the writing stage. The project manager was W. Bolton, assisted by K. Snape.

Self-learning

As an aid to self-learning, questions are included in every chapter. These appear at the end of the chapters with references in the margin of the chapter text (for example Q1.2), indicating the most appropriate position for self-learning use. Answers to each question are given at the back of the book.

The books in this series have therefore been developed for use in either the classroom teaching situation or for self-learning.

Introduction

This book is written to the objectives specified in the TEC Unit Microelectronics Systems U79/603. The main aims of the book are, first, to describe the architecture of a typical microprocessor and the different instruction types normally available with such devices and, secondly, to describe the various system components which comprise a microcomputer and to show how the microcomputer can be used, with the aid of additional interface circuitry, with a range of different peripheral devices.

A currently available microprocessor has been selected for example purposes. This ensures that the microprocessor instructions used are typical of those normally available and avoids the potentially confusing necessity of describing the many small differences between one microprocessor and another.

The book contains many program examples which are used extensively to illustrate the different instruction types and programming principles being introduced. It is intended that these can be run with little or no modification on a suitable manufacturer's single-board prototyping system and hence provide the basis for a suitable supporting laboratory programme.

Acknowledgements

I would like to thank my colleague from the University of Sussex, Phil Cooke, and the series co-ordinator, Norman Bonnett, for their critical reading of the manuscript and helpful suggestions. My thanks also go to Christine Thornton-Clough for typing the draft script and to Peter Hill for his suggestions when preparing the final manuscript.

FRED HALSALL

Chapter 1 Digital computer principles

Objectives of this chapter *When you have completed studying this chapter you should be able to:*

1 *Appreciate the historical developments that have taken place in semiconductor technology since the invention of the transistor.*
2 *Understand the meaning and the resulting advantages of the stored program concept.*
3 *Understand the fundamental mode of operation of a digital computer and the significance of a programmable device.*
4 *Appreciate that all information within a digital computer is stored in a binary-coded form.*
5 *Describe some of the different types of interface circuit used with microcomputer systems.*
6 *Convert a decimal number into its equivalent binary form and vice versa.*
7 *Convert octal and hexadecimal numbers into their equivalent binary forms.*
8 *Convert octal and hexadecimal number into their equivalent binary forms.*
9 *Draw a block diagram of a typical microcomputer and explain the function of each component part.*
10 *Explain how a microprocessor communicates with its memory to execute a stored program.*

1.1 Historical development

The digital computer is the culmination of considerable technological evolution stemming from the days of Charles Babbage and his 'analytical engine'. Babbage is generally credited with the invention of the digital computer although his machine was in fact mechanical and it was not made to function satisfactorily until after his death in 1833.

Progress after this date was slow and it was not until after the Second World War that the first electronic digital computers were built using thermionic valves. The reliability of these machines was low, however, and it was the advent of the transistor in 1947 that heralded the birth of the general-purpose electronic digital computer as we know it today.

The transistor forms the basis of all the electronic circuits from which a digital computer is made and, since its invention in 1947, semi-conductor technology with which transistors are made has evolved very rapidly. It is now commonplace, for example, for an individual to wear on the wrist a digital watch based on an integrated circuit comprised of about a thousand transistors. Similarly, many people now use a small programmable calculator (a limited form of digital computer) which is also constructed using a single integrated circuit containing several thousand transistors.

Semiconductor technology reached a level of development in the early 1970s when it became possible to construct a powerful computer processing unit comprised of 10,000 transistors on a single integrated circuit. This is known as a *microprocessor* which, with some additional integrated circuits, forms a powerful general-purpose digital computing system.

Developments in semiconductor technology are still continuing. Hence, although current technology places a practical limit of about 100,000 transistors per integrated circuit, there is much speculation about possible devices containing one million transistors. The speculation is no longer about if such a device is possible but when it will be available. The historical development of semiconductor technology is illustrated in Table 1.1 and a photomicrograph showing a typical microprocessor which is comprised of 20,000 transistors is shown in Figure 1.1.

1.2   The stored-program concept

Although a simple electronic calculator is able to perform similar arithmetic operations as a digital computer, a calculator relies on a human operator to key-in each operation in the required sequence. A digital computer, however, can perform many millions of operations without the need for human intervention. This is possible because the sequence of operations required by the computer to perform the

Table 1.1   *Historical development of computers and their technology*

Development	Date
Babbage's analytical engine	1833
Invention of thermionic valve	1910
First electronic digital computer	1946
Transistor invented	1947
First transistor digital computer	1960
Integrated circuit invented ($\sim$ 10 transistors)	1963
First microprocessor ($\sim$ 10,000 transistors)	1970
Integrated circuits with $\sim$ 100,000 transistors	1981

Figure 1.1 Photomicrograph of a microprocessor. (*Courtesy* Intel Corporation)

desired task is stored within the computer itself. The computer can therefore perform each operation and step on to the next at a speed determined by the technology with which it is made rather than the very slow speeds at which a human is able to key-in these operations.

The list of operations that performs the desired task is called *a program* and, since this program is stored within the computer, all digital computers are said to operate using the *stored-program concept*.

The stored-program concept has two significant advantages. First, as has been indicated, once the sequence of operations has been entered and stored within the computer, the time taken to perform the complete computation does not depend on the human operator. That is, the computer can complete each operation and move on to the next without human intervention. Secondly, since the sequence of operations is stored within the computer, the computational task performed by the sequence of operations can be performed many times over on new data simply by entering a different set of data values.

1.3 Mode of operation

A digital computer is a very flexible general-purpose machine that can be applied to a wide range of different applications. This is a direct consequence of the general-purpose nature of its design. This means that the same computer made by a manufacturer can be applied to a variety of different applications, because it is designed to be applied to solve a particular set of application requirements *after* it has been made.

The computer provided by a manufacturer is simply a device that can perform a variety of different operations. These include operations to input data into the computer (the data to be processed), operations to process data (perform the required computations) and operations to output data from the computer (the results). The computer is therefore used for a particular application by the user selecting those operations needed to perform the required task (from the list of operations available), ordering these in the appropriate sequence thus forming the program and then storing this program (the selected list of operations) in the computer. The program is then *run* or *executed* to perform the required task. The same computer, therefore, can be readily used for a different application by selecting a different set of operations, i.e. by changing the stored program. This is illustrated in Figure 1.2.

In order to gain an appreciation of the basic operating mode of a digital computer, it is perhaps helpful to first examine the mode of operation of a modern electronic calculator.

A simple electronic calculator can perform a standard set of operations on numbers – addition, subtraction, multiplication and

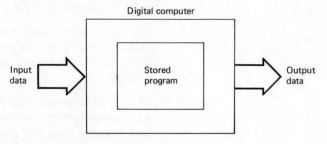

Figure 1.2 Stored program concept

division. A particular task is then solved with such a machine by the human operator pushing keys to specify the sequence of these operations which is required to perform the appropriate computation.

The overall time taken to perform a particular computation using a calculator clearly depends on the speed of the human operator. More advanced electronic calculators therefore contain a *memory* and are then said to be *programmable*. That is, the operator can preload or store within the memory of the calculator the sequence of operations – the program required – to perform the desired computational task.

A programmable calculator can be considered, therefore, as a special-purpose digital computer. It is special-purpose in the sense that it is arranged to perform a particular range of numerical calculations on data entered via a keyboard and it produces an output in the form of data presented on a numerical display. The mode of operation of a programmable calculator, however, is the same as a general-purpose digital computer since both operate using the stored program concept.

As has been mentioned, the concept of the program being stored within the computer is a most important one since it means that the computer can then operate at its own speed and independently of the user. After the list of operations which make up the program has been stored within the computer, the computer operates by obtaining or reading each operation from the store in sequence, decoding or interpreting this operation and then performing or executing the particular function.

The basic operations executed by a computer are known as *machine instructions* and in general a machine instruction specifies both the type of operation and the source of the data on which the operation is to be performed. There are, therefore, instructions to input data values into the computer, instructions to process or manipulate this data (perform arithmetic and other computations) and instructions to output the subsequent results from the computer. The different instruction types provided with a digital computer will be described

Q1.1 in subsequent chapters.

See note in Preface about questions

1.4 Digital circuits

A computer performs operations (such as addition and subtraction) on data that are exact. It is built, however, like any other electronic circuit from components whose values are not *exactly* known.

Any electronic component, even when it is a component within an integrated circuit, has a nominal value for which it was designed. Its actual value, however, depends on the particular manufacturing circumstances associated with that device and tolerances of the order of 5% are typical. How then can a circuit consisting of perhaps many thousands of such devices perform exact operations?

The answer to this question is of fundamental importance to modern electronics. The solution is to use a circuit that operates using just two voltage levels to represent a unit of information rather than using, say, ten absolute voltages. Clearly, if the circuit has to discriminate between just two levels, the particular voltage for each level can have a wide tolerance to indicate the correct level and hence symbol being represented.

The system of representing information using two levels or states is called the *binary (two-state) system* and the two symbols used are written as 1 and 0. This basic unit of information is called a binary digit or *bit*, and a circuit that operates on binary data is called a *digital circuit*.

The most common scheme for representing a binary digit uses the two voltage levels indicated in Figure 1.3. The figure shows that if a voltage from the circuit is in the range 2.4 to 5.0 V then the binary

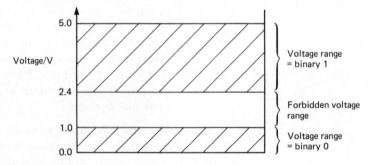

Figure 1.3   Representation of a binary digit in a digital circuit

digit 1 is being represented. If the voltage is in the range 0 to 1.0 V then the binary digit 0 is represented. The only requirement for a circuit to operate exactly using these levels is that the voltage must either be greater than 2.4 V or less than 1.0 V, i.e. it must not produce a voltage in the range 1.0 to 2.4 V which in practice ensures a reliable *margin* between the two levels.

If a computer used just one of these circuits it would, of course, be useless. If, however, a number of similar circuits are used together to represent a group of binary bits, a convenient representation of almost limitless information is possible. This grouping of digital circuits provides the facility of representing a number in the binary system and this provides a convenient and important approach to manipulating data within a digital computer.

1.5   The binary system

Since all information stored and manipulated by a digital computer is in a binary-coded form, before considering the structure and operation of a computer, it is helpful to first gain an understanding of the binary system.

First consider the construction of a typical number in the decimal system. The decimal system uses the ten symbols or digits 0–9 and when writing down a number in the decimal system the value of a particular digit is determined by its position relative to the decimal point. For example, the decimal number 186324 is interpreted as $(1 \times 10^5) + (8 \times 10^4) + (6 \times 10^3) + (3 \times 10^2) + (2 \times 10^1) + (4 \times 10^0)$. Each digit is one of the ten symbols 0–9 and is multiplied by a power of ten determined by the digit's position relative to the decimal point.

Decimal numbers are said to have a *base* of ten because each column differs from the preceding column by a factor of ten. Many other number systems are possible, however, but if the base is chosen as two the system requires just two symbols to represent each digit. The columns then differ by powers of two and hence this is known as the binary number system.

Binary-to-decimal conversion

The binary system uses the two symbols 0 and 1 and hence an example of a binary number is 101101. By comparison with the decimal system it can readily be deduced that this represents $(1 \times 2^5) + (0 \times 2^4) + (1 \times 2^3) + (1 \times 2^2) + (0 \times 2^1) + (1 \times 2^0)$. It is equivalent, therefore, to $(1 \times 32) + (0 \times 16) + (1 \times 8) + (1 \times 4) + (0 \times 2) + (1 \times 1) = 45$ in the decimal system. This is often represented as

$$101101_2 = 45_{10}$$

where the subscripts are used to indicate the base of the appropriate number system.

Example 1.1   Derive the decimal equivalent of 10101100_2.
Answer $10101100_2 = (1 \times 2^7) + (0 \times 2^6) + (1 \times 2^5) + (0 \times 2^4) +$
$(1 \times 2^3) + (1 \times 2^2) + (0 \times 2^1) + (0 \times 2^0)$
$= 128 + 32 + 8 + 4$
$= 172_{10}$

Example 1.2   Derive the decimal equivalent of 1001001101100011_2.
Answer $1001001101100011_2 = (1 \times 2^{15}) + (1 \times 2^{12}) + (1 \times 2^9) +$
$(1 \times 2^8) + (1 \times 2^6) + (1 \times 2^5) + (1 \times 2^1) +$
(1×2^0)
$= 32768 + 4096 + 512 + 256 + 64 + 32 + 2 + 1$
$= 37731_{10}$

The second example illustrates that when performing binary-to-decimal conversions it is useful to have a table of the values of 2^n for a range of numbers 1 to *n*. This is therefore given in Table 1.2 for reference.

Q1.3

Decimal-to-binary conversion

In order to manipulate decimal quantities within a computer it is necessary to first translate them into their binary form. This process can be accomplished by successive division by two and recording the remainders. This is shown in the following example.

Example 1.3   Derive the binary equivalent of 1833_{10}.

Answer   Dividing by 2:

$$\begin{array}{r} 916 \quad \text{remainder} \quad 1 \\ \hline 2)\overline{1833} \end{array}$$

Dividing again:

$$\begin{array}{r} 458 \quad \text{remainder} \quad 0 \\ \hline 2)\overline{916} \end{array}$$

Etc.:

$$\begin{array}{r} 229 \quad \text{remainder} \quad 0 \\ \hline 2)\overline{458} \end{array}$$

$$\begin{array}{r} 114 \quad \text{remainder} \quad 1 \\ \hline 2)\overline{229} \end{array}$$

$$\begin{array}{r} 57 \quad \text{remainder} \quad 0 \\ \hline 2)\overline{114} \end{array}$$

$$\begin{array}{r} 28 \quad \text{remainder} \quad 1 \\ \hline 2)\overline{57} \end{array}$$

$$\begin{array}{r} 14 \quad \text{remainder} \quad 0 \\ \hline 2)\overline{28} \end{array}$$

$$\begin{array}{r} 7 \quad \text{remainder} \quad 0 \\ \hline 2)\overline{14} \end{array}$$

$$\begin{array}{r} 3 \quad \text{remainder} \quad 1 \\ \hline 2)\overline{7} \end{array}$$

$$\begin{array}{r} 1 \quad \text{remainder} \quad 1 \\ \hline 2)\overline{3} \end{array}$$

$$\begin{array}{r} 0 \quad \text{remainder} \quad 1 \\ \hline 2)\overline{1} \end{array}$$

The binary equivalent is then the assembled remainders read from the bottom upwards, i.e. $1833_{10} = 11100101001_2$.

This can be checked by the reverse expansion process, i.e. by finding the decimal equivalent of 11100101001_2:

$$11100101001_2 \ = 2^{10} + 2^9 + 2^8 + 2^5 + 2^3 + 2^0$$

which, using Table 1.2 gives:

$$11100101001_2 \ = 1024 + 512 + 256 + 32 + 8 + 1 \ = 1833_{10}$$

Table 1.2   *Powers of 2*

n	2^n
0	1
1	2
2	4
3	8
4	16
5	32
6	64
7	128
8	256
9	512
10	1024
11	2048
12	4096
13	8192
14	16384
15	32768
16	65536
17	131072
18	262144
19	524288
20	1048576
21	2097152
22	4194304
23	8388608

The foregoing shows that the binary system provides a means of representing any number using only two symbols. Hence, as was described in the previous section, it is possible to represent numbers accurately within a computer by using a group of digital circuits in which each circuit represents a singly binary digit. In addition to using the binary system to represent numbers within the computer, however, the binary system is also used to represent the instructions which are stored within the computer memory. The actual form of representation will be described in a later chapter but as a result, when examining the operation of a microcomputer system, binary patterns are always being considered. This can be very tedious and, since binary numbers are often comprised of a large number of bits, they are prone to errors when communicating this information to others.

Q1.2

It is for this reason that an alternative shorthand form of representation is used. The method is to group a number of bits together and then represent that group with an equivalent number or character. The two most frequently used number of bits are 3 and 4 which correspond to the octal and the hexadecimal number systems, respectively.

The octal system

The octal number system, as the name suggests, has a base of eight and uses the eight symbols 0–7. Octal numbers, however, are rarely used in their own right but, as was indicated above, are often used as a shorthand form of representing a binary number. This is possible because the base of the octal system (8) is a power of two. There is, therefore, a very simple conversion procedure between binary and octal and between octal and binary.

Consider the powers of two of the columns of a binary number:

$$2^8 \; 2^7 \; 2^6 \; 2^5 \; 2^4 \; 2^3 \; 2^2 \; 2^1 \; 2^0$$

If these are divided into groups of three bits the sum of each group is the same as the maximum value of an octal digit:

$$\underbrace{2^8 \; 2^7 \; 2^6} \qquad \underbrace{2^5 \; 2^4 \; 2^3} \qquad \underbrace{2^2 \; 2^1 \; 2^0}$$

$$= 448 \qquad\quad =: 56 \qquad\quad = 7$$
$$= 7 \times 8^2 \qquad = 7 \times 8^1 \qquad = 7 \times 8^0$$

Hence in order to obtain an octal number representation of a binary number it is only necessary to divide the binary number into groups of three bits and perform the conversion of each three-bit group into the equivalent octal digit. For example,

$$101111001_2$$

can be represented as an octal number by first dividing the binary number into three-bit groups

$$1\,0\,1 \quad 1\,1\,1 \quad 0\,0\,1$$

Table 1.3 *Some binary and their equivalent octal numbers*

and then performing the binary-to-octal conversion of each three-bit group, i.e.

$$\underbrace{1\,0\,1}_{5} \quad \underbrace{1\,1\,1}_{7} \quad \underbrace{0\,0\,1}_{1}{}_2 = 571_8$$

Binary	Octal
0	0
1	1
10	2
11	3
100	4
101	5
110	6
111	7
1000	10
1001	11
1010	12
1011	13
1100	14
1101	15
1110	16
1111	17
10000	20

Thus the octal shorthand for 101111001_2 is 571_8 – a rather easier number to remember.

Conversion from octal to binary is simply the reverse procedure. For example, 571_8 is converted by first representing each octal digit in its three-bit binary equivalent form, and then grouping the resulting bits together:

$$
\begin{array}{cccccc}
5 & 7 & 1_8 & & = & 101\ 111\ 001_2 \\
101 & 111 & 001 & & = & 5\quad 7\quad 1
\end{array}
$$

That is, $571_8 = 101111001_2$. For reference, Table 1.3 gives some binary numbers and their equivalent octal forms.

The octal system is a particularly attractive shorthand for representing binary numbers that have multiples of three bits, such as 9 or 12. Many computers, however, utilise multiples of four bits such as 4, 8, 16 and 32. For these machines, therefore, the octal system is a little clumsy and an alternative system which groups four bits has obvious advantages. This corresponds to a number system which has a base of 16 and is called the hexadecimal system.

Q1.4

The hexadecimal system

The hexadecimal system has a base of sixteen and hence sixteen symbols are necessary to represent each digit. The decimal system, however, provides only ten single digit symbols (0–9) so a further six symbols are required. The sixteen symbols used are the ten decimal digits 0–9 plus the six alphabetic characters A, B, C, D, E and F. The alphabetic characters are used to represent the six numbers that in the decimal system would require two digits, i.e. 10, 11, 12, 13, 14 and 15, respectively. A typical four-digit hexadecimal number is therefore $9A7F_{16}$.

Since hexadecimal numbers have a base of sixteen, it is possible to convert from binary to hexadecimal by grouping together four binary bits and representing them with a single hexadecimal symbol. Table 1.4 shows the sixteen hexadecimal symbols and their equivalent binary numbers.

For example,

$$\underbrace{1\ 0\ 1\ 1}_{B}\ \underbrace{0\ 1\ 1\ 1}_{7}{}_{2} = B7_{16}$$

Similarly,

$$\underbrace{1\ 1\ 0\ 0}_{C}\ \underbrace{0\ 1\ 0\ 1}_{5}\ \underbrace{1\ 0\ 0\ 1}_{9}\ \underbrace{1\ 1\ 1\ 1}_{F}{}_{2} = C59F_{16}$$

Converting the hexadecimal representation of numbers to binary is simply the reverse procedure. For example,

$$\begin{array}{cccc} C & 5 & 9 & F \\ 1\ 1\ 0\ 0 & 0\ 1\ 0\ 1 & 1\ 0\ 0\ 1 & 1\ 1\ 1\ 1 \end{array} = 1100010110011111_{2}$$

Similarly,

$$\begin{array}{cccc} 2 & A & F & 1 \\ 0\ 0\ 1\ 0 & 1\ 0\ 1\ 0 & 1\ 1\ 1\ 1 & 0\ 0\ 0\ 1 \end{array} = 0010101011110001_{2}$$

It is quite common to see hexadecimal numbers written as C59FH (where the H stands for hexadecimal) or C59F (hex) to represent $C59F_{16}$. Since many microcomputers use a basic unit of either 8 or 16 bits, the hexadecimal system is a very popular shorthand for representing binary patterns that arise within these systems and will be used extensively throughout this book.

Q1.5

Table 1.4 *Four-bit binary numbers and their equivalent hexadecimal digits*

Binary	Hexadecimal
0000	0
0001	1
0010	2
0011	3
0100	4
0101	5
0110	6
0111	7
1000	8
1001	9
1010	A
1011	B
1100	C
1101	D
1110	E
1111	F

1.6 The computer interface

Although the input data and the subsequent results produced by an electronic calculator are always numeric, i.e. the data represents decimal numbers, the input and output data used by a digital computer may represent, in addition to the normal numeric values, a variety of different parameters.

In an application in which a computer is being used to control the temperature in a room, for example, the input data may indicate the current temperature of the room and the output data the quantity of heat to be produced by a heater. Similarly, in an application where a computer is being used to control the actions of a washing machine, the input data may indicate such parameters as the temperature and the level of the water in the tub, and the output data signals to switch the wash motor or water pump on or off and so on.

Irrespective of the application, however, the data input and output by the computer itself are always in a binary-coded form and hence the same operations can be performed on this data irrespective of whether the data represents a numeric value or, say, the temperature in a room. Indeed, the computer has no knowledge of what the data represents and it is only the user who manipulates and interprets the data as representing a specific value depending on the application.

This is accomplished by providing suitable *interface circuits* between the computer and the input and output devices being used. These circuits effectively isolate the computer from the specific input and output devices since, irrespective of the device, they always present a standard binary-coded interface to the computer.

For example, if the input data is a continuously varying (analogue) signal from a temperature-sensing device representing, say, the temperature of a process, then an *analogue-to-digital converter (ADC)* is used to produce the digital (binary-coded) equivalent of the temperature. Similarly, if an analogue signal is required by the output device a *digital-to-analogue converter (DAC)* is used to convert the digital (binary-coded) output produced by the computer into its equivalent analogue value.

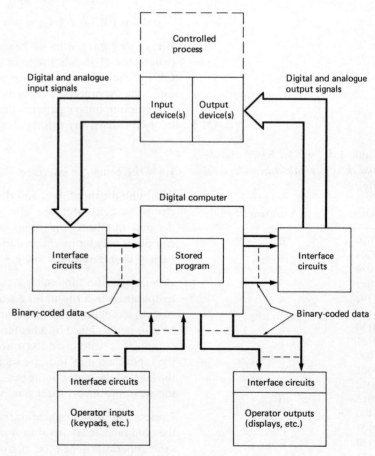

Figure 1.4 The computer interface

If the input data is already in a digital form then this may be read directly. For example, the state of a level switch (on or off) or the state of a control valve (open or closed) can both be represented by the state of a single binary digit – 1 to indicate the device is on (closed) and 0 to indicate the device is off (open). Similarly, a single binary digit output by a computer can be used to switch a relay on or off – 1 to switch it on, 0 to switch it off – which in turn could control the switching (on or off) of a large motor. This is shown diagrammatically in Figure 1.4 and some examples of typical interface circuits used with digital computers are listed in Table 1.5.

Table 1.5 *Some examples of interface circuits used with a digital computer*

Input source	Interface circuit
Two-state device – level switch, thermostat, etc.	Can be input directly
Decimal digit (0–9) from a simple numeric keypad	Digital *encoding* circuit which produces the 4-bit binary-coded equivalent of the decimal digit selected
Hexadecimal digit (0–9 plus A–F) from a hexadecimal keypad	Digital encoding circuit which produces the 4-bit binary-coded equivalent of the hexadecimal digit selected
Alphanumeric character (*alpha*betic A–Z or *numeric* 0–9) from a full keyboard	Digital encoding circuit which produces the 8-bit binary-coded equivalent of the selected characters
Analogue signal (continuously varying), e.g. from a pressure or temperature sensing device	Analogue-to-digital converter (ADC)

Required output	Interface circuit
Two-state device – relay, control valve, etc.	Can be controlled directly
Numeric or hexadecimal display	*Decoding* circuit which produces the drive signals for the (seven-segment) display
Alphanumeric character display	Decoding circuit which produces the drive signals to display the character on a visual display unit (VDU) screen
Analogue signal, e.g. a voltage level	Digital-to-analogue converter (DAC)

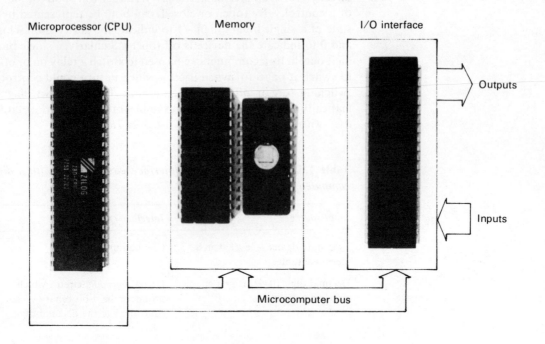

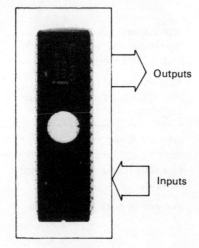

Figure 1.5 Microcomputer integrated circuits: (*a*) a microprocessor, memory and I/O interface circuit (actual size); (*b*) a single-chip microcomputer

1.7 Basic structure

As has been indicated, both the source of the input data and the type of output data required depends on the application to which the computer is being put. All data within the computer itself, however, and also the *actual* data input and output by the computer, are in a binary-coded form. The input and output instructions available with a computer, therefore, simply input or output a fixed length (usually 8-bit) binary-coded value from or to an *interface port* of the computer. The latter are so called because they form the interface between the computer and the interface circuits discussed in the previous section.

A digital computer is comprised, therefore, of three component parts: a *memory unit* which is used to hold both the list of instructions which make up the program and any data values which are awaiting processing, a *central processing unit (CPU)* which interprets and executes each program instruction and an *input-output (I/O) interface unit* which provides one or more input and output ports.

If the central processing unit is made as a single integrated circuit, it is called a *microprocessor* and the complete system – microprocessor, memory and I/O interface – is then called a *microcomputer*. As an example, the three integrated circuits which make up a currently available microcomputer are shown in Figure 1.5(a). Indeed, integrated circuit technology has now reached the stage where it is possible to have a microprocessor and a limited amount of memory and I/O capability on a single integrated circuit and an example of a 'single-chip' microcomputer is shown in Figure 1.5(b).

Clearly, the three units mentioned above must be able to communicate with each other, for example, to fetch each instruction from the memory unit to the microprocessor or to input or output a data value to or from the microprocessor and the I/O interface unit. This is accomplished using an electrical interconnection system called the *computer bus* or *highway* since, as will be seen in the next chapter, *all* communications within the computer are carried out using the
Q1.6   same (bussed) electrical signal lines. This is illustrated in Figure 1.6.

1.8 The fetch–execute cycle

Once the list of instructions which comprise a program has been evolved to solve a particular task, the complete list of instructions is loaded into the microcomputer memory and the program is then run or executed.

During program execution the microprocessor fetches each instruction from the memory in sequence – via the computer bus – determines the action specified in the instruction and performs or

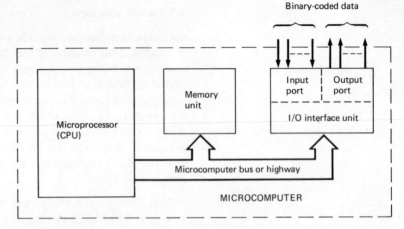

Figure 1.6 Basic structure of a microcomputer

executes this action. The microprocessor is said to operate, therefore, in a two-phase mode: during the first phase it fetches the instruction from the memory – the *fetch phase* – and during the second phase it executes the action specified in the instruction – the *execute phase*. The complete fetch–execute cycle is then known as an *instruction cycle* and this is illustrated in Figure 1.7.

As will be seen in Chapter 3, some microprocessor instructions require a single *byte* (8 bits) of memory space to store the instruction whilst some require two bytes and some three bytes. During the fetch phase of each instruction cycle, however, the microprocessor can determine from the first byte of the instruction both the number (if

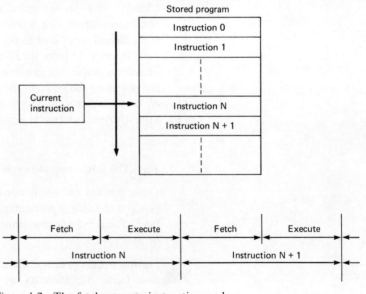

Figure 1.7 The fetch–execute instruction cycle

any) and meaning of the remaining bytes. It can therefore fetch the requisite number of bytes which make up the complete instruction during the fetch phase prior to executing the instruction.

Questions

1.1 Explain the meaning of the following terms:
(a) the stored-program concept
(b) a programmable device.

1.2 Convert the following decimal numbers into their equivalent binary form: 27, 63, 226.

1.3 Convert the following binary numbers into their equivalent decimal forms: 00101101, 10101010, 11011011.

1.4 Represent the following binary numbers in their octal form: 101011, 111001110, 100010110000.

1.5 Represent the following hexadecimal numbers in their equivalent binary form: C2, 8E1, 3B2F.

1.6 Draw a block diagram of a typical microcomputer and explain the function of each component part. Describe the mechanism by which these components communicate and the way the microprocessor executes a program stored in its memory.

Chapter 2 Microcomputer architecture

Objectives of this chapter *When you have completed studying this chapter you should be able to:*

1 *Draw a block schematic diagram and explain the operation of the three component parts which comprise a microcomputer: the memory unit, the microprocessor and the I/O interface unit.*
2 *Understand the term memory map and the difference between the terms RAM, ROM, UVPROM and EAROM.*
3 *Understand the structure of the computer bus and explain how information is transferred between the units which comprise the microcomputer using timing diagrams.*

2.1 Introduction

The three fundamental units of a microcomputer were introduced in the previous chapter. They are the memory unit which is used to store the program and also any data values which are associated with it, the microprocessor (CPU) which interprets and executes the list of instructions which comprise the program, and the I/O interface unit which provides one or more input and output ports to connect the microcomputer to the particular input and output devices being used for the application. This chapter describes the structure and operation of these units in detail and also the structure of the computer bus which is used to interconnect these units.

2.2 The memory unit

The memory unit can be considered as being comprised of a list of separately identifiable boxes called *locations*. Each location can hold or *store* a single fixed-length binary pattern which, for the majority of microprocessor systems, is comprised of 8 bits or *1 byte*. The 8-bit value currently stored in a location is known as the *contents* of that location.

Each location is identified by means of a separate number which is known as the *address* of that location. Although the number of locations (and hence addresses) required for many microprocessor applications is small – typically less than 2,000 – for other more sophisticated applications quite large numbers of locations are required. To allow for a range of applications, therefore, the number

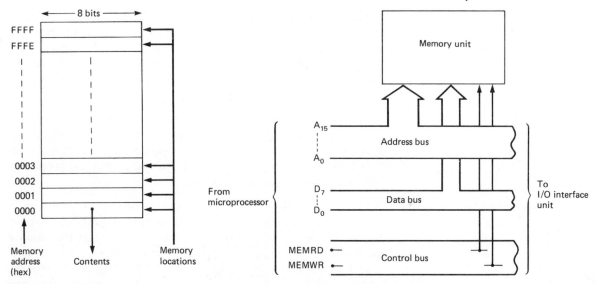

Figure 2.1 The memory unit

Figure 2.2 Communication with the memory unit

of locations available with a microcomputer memory can normally be expanded up to 65,536 (decimal).

The first location in the memory is always assigned the address zero and hence the range of addresses required to identify each location in a microcomputer memory is therefore from 0 to 65,535. In the binary system the decimal number 65,535 corresponds to the binary number 1111111111111111, i.e. sixteen binary ones. In the hexadecimal system this corresponds to the four hexadecimal characters FFFF and hence the range of memory addresses available with a microcomputer system is normally expressed as 0000 (hex) → FFFF (hex). This is shown diagrammatically in Figure 2.1.

Q2.1

To gain access to the contents of a memory location, the address of the location is presented to the memory unit on the sixteen binary *address lines* (A_0 to A_{15}) and a signal given to initiate the access or *read* operation. The latter is called a *memory read* signal (MEMRD) and it is placed on the memory read *control line*. The memory unit then responds by placing the contents of that location (an 8-bit value) on the eight *data lines* (D_0 to D_7).

Similarly, to put a value into a memory location the address of the location is presented on the sixteen address lines, the 8-bit value to be stored is placed on the data lines and a signal given to initiate the storing or *writing* operation. The latter is therefore called the *memory write* signal (MEMWR) and is in turn placed on the memory write control line. This is illustrated in Figure 2.2.

The figure shows the appropriate set of lines used for communicating with the memory unit to read and write a value from or to a location within the memory. As will be seen, the same address, data and control lines used for communicating with the memory unit are also used for communicating with the I/O interface unit. The appropriate lines are therefore known as the *address bus*, the *data bus* and the *control bus*, respectively, and collectively form the computer bus or highway mentioned in Chapter 1.

It should be noted that the data bus during a read operation holds a value to be *input* by the microprocessor and during a write operation it holds a value *output* by the microprocessor. The data bus is therefore referred to as a *bidirectional bus* and its electrical properties will be discussed in a later chapter.

A typical set of electrical or *timing signals* used to read and write a value from and to a memory location is shown in Figure 2.3. They are known as *timing diagrams* since the waveforms (binary transitions) are shown in time sequence with each other. Since each address and data line will either be a binary 1 or 0 depending on the specific address and data values being used, the address and data lines are shown in the figure as being either high (binary 1) or low (binary 0).

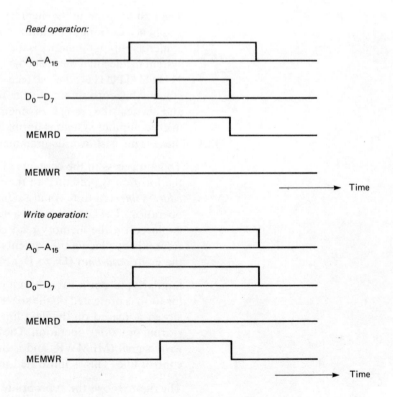

Figure 2.3   Timing signals for read and write operations

Read-only and random access memory

For many dedicated microcomputer applications, i.e. applications in which the stored program remains fixed, the memory unit is normally comprised of two segments:

1   *A read-only memory* section which is used to hold or store the fixed, non-changing list of machine instructions which comprise the program.

2   A *random access memory* section which is used to hold any data values associated with the program, e.g. data values awaiting further processing or waiting to be output to a port within the I/O interface unit.

Read-only memory or ROM has the property that once the information has been fixed into it during its manufacture its contents cannot be changed and consequently it can only be operated in a read-only mode. It also has the property of being *non-volatile* since once its contents have been fixed at manufacture, the same contents will be maintained each time the electrical supply is applied to it. Thus for a dedicated application, the same program will be stored within its memory each time the electrical supply is applied to the micro-computer.

Random access memory or RAM has the property that, in addition to being able to read the contents of a location, new information can be written into a location. It is therefore also referred to as read/write memory. RAM is normally *volatile* since although its contents will remain unchanged all the time electrical power is supplied to it, once the supply is switched off its contents will be lost and the memory will contain random values when the power is reapplied.

The binary patterns representing the program instructions which are fixed into a ROM during its manufacture cannot be changed and hence it is essential that the program is correct and free from errors prior to its manufacture. Also, the cost of producing a factory-made ROM is high if only a small number of such devices (that is, devices with the same program) are required. Factory-made ROMs therefore are only used in large-volume applications such as washing machine controllers and computer games.

A more economic solution for those applications which require only small numbers of similar devices or the facility for changing the program stored within it after its manufacture (for example to add an additional feature or to correct an error), is to use an *erasable programmable ROM or EPROM*.

During its normal operating mode an EPROM behaves exactly the same as a factory-made ROM. It has the property, however, that the memory pattern stored within it can be changed by the user in a controlled manner. They are particularly useful for prototyping

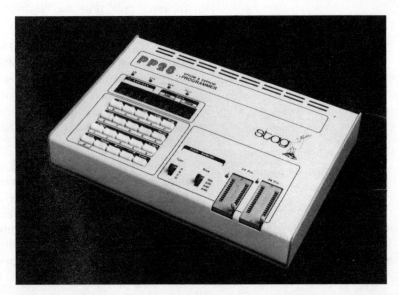

Figure 2.4 UV EPROM *(top)* and PROM programmer *(bottom)*. *(Courtesy: Stag Electronic Designs Ltd)*

systems, therefore, in which the program may need to be changed as the product is being developed.

The contents of an EPROM are erased either by exposure to intensive ultra-violet light through a 'window' on the integrated circuit itself – *UV EPROM* – or by applying a voltage to specific connections (pins) on the integrated circuit – electrically alterable ROM or *EAROM*. The new contents are then written or stored in the device using a special piece of equipment called a *PROM programmer*. An example of a UV PROM and an associated PROM programmer is shown in Figure 2.4. It should be stressed, however, that during their normal operating mode EPROMs behave exactly the same as factory-made ROMs.

The memory map

Although the total number of locations or *address space* provided with a microcomputer can be large, for many small dedicated applications the actual quantity of memory (read-only and random access) can be quite small. The range of addresses that are used for a particular application and the type of memory devices in each range is indicated by means of a *memory map* of the system.

An example of a memory map for a small but typical microcomputer application is shown in Figure 2.5. In this application it is assumed that the fixed program memory requires up to 2,048 [0000 to 07FF (hex)] bytes of ROM and up to 256 [2000 to 20FF (hex)] bytes of RAM are required for the associated data.

It is normal to use a disjoint address space in the memory map, i.e. to leave unused locations between each memory type, first to allow for any future expansion and secondly, as will be seen later, to simplify the electronic circuitry required to determine which type of memory is being selected during a read or write operation.

The actual quantity of memory used for the program in a micro-computer application is normally expressed in multiples of 1,024 bytes. This is the basic unit size of an integrated circuit read-only memory and is designated as 1 Kbyte. Hence the above application uses 2 Kbytes of ROM and the maximum address space available with most microprocessors is 64 Kbytes since this corresponds to 64 × 1,024 = 65,536 decimal locations.

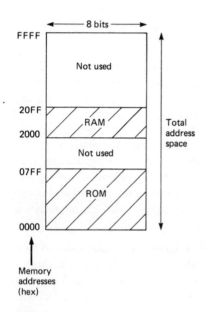

Figure 2.5 A memory map of a system

Q2.2, 2.3

2.3 The microprocessor

The microprocessor executes the program stored within its memory by reading each instruction in sequence. Each instruction is first interpreted or *decoded* to determine the actions necessary to execute this instruction and the appropriate actions carried out.

During program execution both individual program instructions and any intermediate (temporary) data values associated with the program are stored within the microprocessor in a number of special locations called *registers*. A microprocessor register is analogous to a location within the memory unit since a register can hold or store a single binary word either of eight or sixteen bits. A microprocessor register, however, is referred to by means of a symbolic name rather than by an address since each register is used for a specific function.

As has been stated, the microprocessor executes a program stored within its memory by reading each program instruction sequentially. It is therefore necessary for the microprocessor to remember, or keep a record of, the address within the memory unit of the location which holds the next instruction to be executed. The microprocessor thus contains a register called the *program counter* (PC) which is used for this function. Since the program counter holds a memory address it is comprised of sixteen bits.

In addition, the microprocessor contains a register called the *instruction register* (IR) which is used to hold an instruction whilst it is being decoded to determine the actions necessary to execute the instruction. The decoding operation is performed by the *instruction decode unit*. Essentially the latter determines which instruction has been read and, in conjunction with electrical timing signals generated from the *system clock*, produces timing and control signals similar to those shown earlier in Figure 2.3. Collectively the instruction register and the timing and decode unit are known as the *control unit* of the microprocessor since all the electrical signals used throughout the microcomputer are generated by it.

In addition to the basic arithmetic operations of addition and subtraction, a microprocessor can also perform logical operations between two data values. The specific meaning and usefulness of these instructions will be described in a later chapter but the significance here is that the unit within the microprocessor which performs these data-manipulation functions is known as the *arithmetic logic unit* or *ALU*.

The ALU performs the appropriate arithmetic or logical operation (as specified in the instruction being executed) on two values, one of which is always the current contents of a special microprocessor register called the *A register*. The other value is either the contents of a specified memory location or the current contents of one of a *set of registers* which also form part of the microprocessor. The latter are used for holding temporary values awaiting further processing and are normally called register B, register C, etc. The actual number of registers available varies from one microprocessor to another.

The result of an arithmetic or logical operation is normally placed in the A register and hence replaces or overwrites its existing contents. The A register, therefore, is also known as the *accumulator* since its contents can be considered as accumulating the overall result of a specific arithmetic computation.

Finally, as has just been indicated, many microprocessor instructions specify the address of a memory location within them, e.g. the address of the location in memory which holds the second value involved in an arithmetic or logical operation. A second 16-bit register (or a pair of 8-bit registers) is therefore used to hold this address during the execution of instructions that specify a memory address. It is known as the *memory address register* or MAR.

A schematic diagram of a typical microprocessor is therefore as shown in Figure 2.6. It is comprised of a register unit which contains the various registers outlined above, an ALU which performs the various arithmetic and logic operations and a control unit which generates the timing signals to control the flow of information (both instructions and data) between the various units which comprise the microcomputer and between the registers and the ALU within the microprocessor itself.

Q2.4

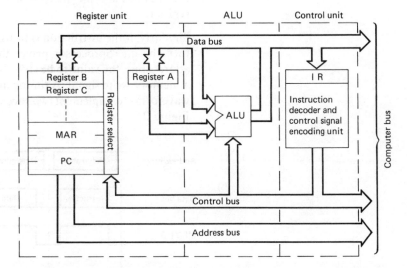

Figure 2.6 Block schematic of a microprocessor

2.4 The computer bus

It is perhaps helpful at this point to consider the timing and other signals on the computer bus generated by the microprocessor to implement some typical machine instructions. To execute a program, the microprocessor operates in a two-phase or fetch–execute mode: during the fetch phase the next sequential instruction is read from memory and during the execute phase the microprocessor carries out the actions necessary to implement the instruction. Each complete fetch–execute cycle is also known therefore as an instruction cycle.

The program counter within the microprocessor always holds in memory the address of the next instruction to be executed. During the fetch phase of each instruction cycle, therefore, the microprocessor places the contents of the program counter on the address bus and generates a memory read signal. The contents of the addressed location (the instruction) are then placed on the data bus by the memory unit and are loaded into the instruction register by the microprocessor ready for decoding and subsequent execution.

The instruction read from memory is then decoded and the appropriate actions carried out to implement the instruction during the execute phase. For example, if the instruction is to read a value from a memory location into, say, a microprocessor register, the microprocessor places the address of the memory location on the address bus and generates a memory read signal. The memory unit then responds by placing the contents of the addressed location on the data bus and the microprocessor loads this into the appropriate register.

Similarly, if the instruction is to write a value into a memory location then the microprocessor places the address of the location on the address bus, the value to be stored on the data bus and then generates a memory write signal. A timing diagram to illustrate a typical set of waveforms to implement two successive instruction cycles is shown in Figure 2.7.

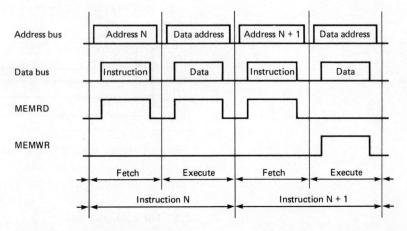

Figure 2.7   Two instruction cycles

The figure illustrates the execution firstly of an instruction which involves a memory read operation and secondly an instruction which involves a memory write operation. During each fetch phase the address output by the microprocessor is the current contents of the program counter and hence these are shown as address N and address N + 1, respectively. Thus the contents of the program counter are always incremented to point to the start of the next instruction in memory after each fetch phase.

As will be seen in a later chapter, some microprocessor instructions require more than a single byte to specify the instruction. For example, some instructions contain a memory address or indeed an actual data value within them. During the fetch phase of the instruction cycle, therefore, the microprocessor must fetch the appropriate number of bytes from memory before starting to execute the instruction.

This is readily accomplished since the microprocessor can deduce from the first byte of each instruction, i.e. the byte loaded into the instruction register, both the number (if any) and meaning of the remaining bytes to be brought from memory. Hence if two further bytes are required, e.g. for a 3-byte instruction, then these are both read from successive memory locations during the fetch phase using two separate read commands. The additional bytes are then loaded into the appropriate microprocessor registers depending on the instruction, e.g. the memory address register if the two bytes form a memory address. Irrespective of the number of bytes in each instruction, therefore, the program counter is always incremented by the
Q2.5   appropriate amount (1, 2 or 3) during each fetch phase.

2.5   The I/O interface unit

The I/O interface unit provides one or more input and output ports to connect the microcomputer to the input and output devices being used for the application. Since the latter are effectively connected to the outside or periphery of the microcomputer they are often referred to as *peripheral devices*.

As was explained in Chapter 1, irrespective of the type of input and output devices being used for the application, the microprocessor simply either inputs (reads) a binary value (usually 8 bits) from an input port or outputs (writes) a binary value to an output port. This value is then converted into a suitable form for connecting to the actual input and output devices by additional interface circuits.

The I/O interface unit is connected to the microprocessor using the same computer bus as is used for connecting the memory unit to the microprocessor. It is therefore necessary for the microprocessor to be able to both write and read data to and from each unit separately. In addition, if more than one input and/or output port is required it is necessary for the microprocessor to be able to select the specific port involved in each input and output operation.

This is accomplished by the microprocessor using part of the address bus to indicate the specific port required in a transfer and also by the microprocessor control unit generating two additional control signals whenever an input or an output instruction is being executed.

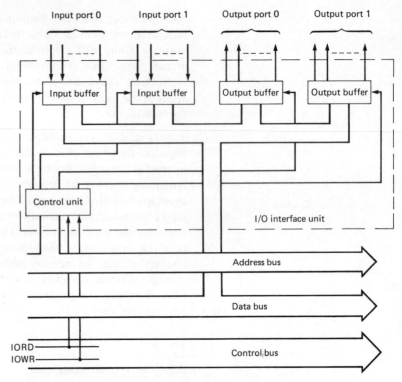

Figure 2.8   I/O interface unit schematic

These are called IORD and IOWR, respectively, and are placed on the control bus by the microprocessor during the execute phase of the instruction cycle. They enable the memory unit and the I/O interface unit to discriminate between memory and I/O read and write operations on the bus.

Each port within the I/O interface unit contains a *port buffer* which effectively isolates the computer bus from any input or output activity being performed by the controlled peripheral devices. An *output buffer* is an electrical circuit which is used to hold or *latch* a data value output by the microprocessor prior to it being accepted by an output device and an *input buffer* is used to hold a data value produced by a device awaiting input by the microprocessor.

Thus to read a value from an input port, the microprocessor simply places the address of the appropriate *port* on the address bus and generates an IORD signal. The I/O interface unit then responds by placing or *enabling* the contents of the addressed port input buffer onto the data bus and this in turn is loaded into the accumulator.

Similarly, to write a value to an output port, the microprocessor places the port address on the address bus, the data to be output on the data bus – usually the current contents of the accumulator – and

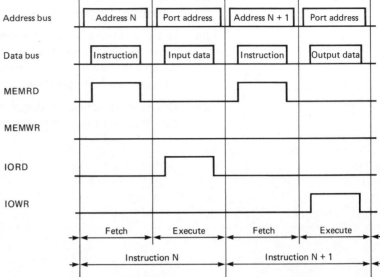

Figure 2.9 I/O timing diagram

generates an IOWR signal. The I/O interface unit then responds by latching the value on the data bus into the output buffer associated with the addressed output port to await being taken by the output device.

A schematic diagram of an I/O interface circuit containing two input ports and two output ports is shown in Figure 2.8 and a timing diagram illustrating a typical input (IN) and output (OUT) instruction is shown in Figure 2.9.

The control unit shown in Figure 2.8 determines from the address bus the specific port involved in the transfer – *address decoding* – and also performs the enabling and latching functions to or from the data bus.

Q2.6

Questions

2.1 Determine the maximum memory space of an 8-bit microprocessor which uses a 12-bit address word. Express your answer in Kbytes.

2.2 A microcomputer system requires 8 Kbytes of PROM and 256 bytes of RAM. Assign suitable start and end addresses for each memory block using disjoint addresses. Assume the microprocessor uses 16-bit addresses.

2.3 A microcomputer uses the following address map:
    0000 → 07FF ROM
    2000 → 23FF RAM
Determine the amount of ROM and RAM memory in the system.

2.4   Draw a block schematic diagram of a typical microprocessor and explain the function of each component part.

2.5   Draw a timing diagram showing the electrical signals on the computer bus during the execution of two successive instructions. Assume the first instruction is a memory reference instruction and the second is an I/O operation.

2.6   Draw a block schematic diagram of an I/O interface unit and explain the function of each component part.

Chapter 3 Microprocessor instructions

Objectives of this chapter *When you have completed studying this chapter you should be able to:*

1. *Understand that a machine-level instruction is made up of two parts: an operation code and an operand part.*
2. *Know that the operand part of an instruction may specify either a specific value or the location of the value on which the operation is to be performed.*
3. *Explain, with the aid of examples, the meaning of the terms: symbolic notation, operation mnemonic, symbolic address.*
4. *Give an example and describe the action of an instruction from each of the four different groups which comprise a typical instruction set.*
5. *Understand the four types of addressing mode used with micro-computer systems.*
6. *Understand and write short programs to illustrate how data is moved (transferred) between the various microprocessor registers and between a microprocessor register and a location within the memory.*
7. *Use a manufacturer's coding sheet to translate a program written in symbolic notation form into its hexadecimal machine code form.*
8. *Appreciate the functions provided with a manufacturer's single-board microcomputer system.*

3.1 Introduction

In the previous chapter it was seen that a microprocessor, like any digital computer, operates using the stored-program concept, i.e. it executes a program that is stored within its memory. The program is comprised of a list of basic binary-coded machine instructions which are executed sequentially by the microprocessor.

The instructions which comprise a program are those selected by the user or *programmer* from the set of instructions available with the chosen microprocessor to perform the required application function or task. Before a machine-level program can be written, therefore, it is first necessary to understand the type of instructions normally available with a microprocessor and the actions performed by these instructions.

Typically, instructions are provided to input and output a binary value to or from an I/O interface port; to move or transfer data between the various components of the microcomputer and between the various registers within the microprocessor; to manipulate data, i.e. perform arithmetic and logical operations on it, and, finally, to control the flow of program execution.

This chapter describes the structure of a machine-level instruction and presents examples of the different instruction types and addressing modes normally used with microprocessor systems. Finally, the facilities provided by manufacturers for developing and testing short machine-level programs are described.

3.2   A machine-level instruction

All machine-level instructions executed by a microprocessor perform a specific function or operation. Typically, this operation may relate to an item of data defined within the instruction or it may relate to an item of data stored within either one of the registers of the microprocessor or one of the storage locations of the memory unit. The majority of instructions, therefore, consist of two different segments or parts: one to specify the specific operation to be performed – *the operation code* (or *opcode*) – and the other to specify either the data value or the location of the data on which this operation is to be carried out – *the operand part*.

Since the latter frequently relates to a memory location, the second part of an instruction is also referred to as *the address part*. The structure of a machine instruction can be shown diagramatically as follows:

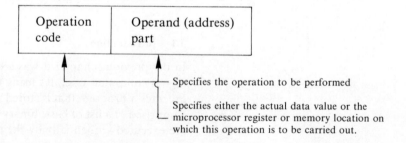

The instructions which comprise a machine-level program are stored within the memory prior to execution. Since the number of *binary digits* (bits) which comprise each memory location is fixed – typically 8 bits (or a byte) for microprocessor systems – the number of bytes required to store each instruction varies for the different instruction types. For example, an instruction that specifies a register in the operand part usually requires only a single byte, whereas an instruction that specifies the data value within it needs two bytes or, if the operand part specifies a memory location, three bytes are normally required.

Irrespective of the number of bytes in the instruction, however, during the fetch phase of the instruction cycle the microprocessor simply reads the first byte of the instruction and deduces from this both the operation to be performed and the number (if any) and meaning of the following bytes which make up the instruction.

When the operand part of an instruction is an address, three separate addresses are implied: two to specify the location of the values to be manipulated – *the source address* – and the third to specify the location of where the result is to be stored – *the destination address:*

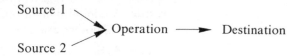

In order to reduce the number of addresses required, therefore, in most systems if an instruction requires two source addresses, the destination address is made the same as one of the source addresses. Moreover, it is often *implied* in the operation code that this address is one of the internal registers – for instance the accumulator or register A.

Thus when the operand part of an instruction specifies an address, only a single address is normally required and the second address (source or destination) is implied in the operation code. With this approach a significant saving in memory space is obtained to store each instruction and hence the complete program.

3.3   Symbolic notation

Although when stored within the memory of the microcomputer a machine instruction is in a binary-coded form, it is clearly extremely tedious and time-consuming to represent an instruction in this form on paper. A far more convenient method is to represent each instruction in its *symbolic notation* form.

Using symbolic notation, a machine instruction is represented as follows:

OPERATION MNEMONIC   SYMBOLIC ADDRESS

The *operation mnemonic* is an abbreviated name which represents the operation code part of the instruction (and hence the order to be performed) and the *symbolic address* the operand or address part. For example,

```
ADD   A,B
```

Symbolic address

Operation mnemonic

may be used to represent the binary-coded instruction

| 10000000 | = 80 (hex)

Typically this instruction is used to add the current contents of one microprocessor register (register B in this example) to the contents of another (register A). Notice that in this example the instruction specifies the two registers involved but in many microprocessors register A is implied in the operation mnemonic and only register B is specified. Another example,

```
LD   A,B
```

Symbolic address

Operation mnemonic

may be used to represent the binary-coded instruction

| 01111111 | = 7F (hex)

Typically this instruction is used to load (move) the current contents of one internal microprocessor register (register B in this example) to another (register A). Similarly,

```
LD   (20A0),A
```

Symbolic address

Operation mnemonic

may be used to represent the binary-coded instruction

| 00110010 | = 32 (hex) } LD
| 10100000 | = A0 (hex) }
| 00100000 | = 20 (hex) } 20A0

Typically this instruction is used to load (store) the current contents of a microprocessor register (register A) into a location within the memory unit [20A0 (hex)]. Notice in this example that the symbolic address specifies the actual memory address in hexadecimal form (20A0). The effect of the instruction is therefore as shown in Figure 3.1. It is assumed that the initial contents of register A are FE (hex) – the result, say, of a previous instruction.

It can readily be seen from the above examples that when an instruction is written in its symbolic notation form, it is far more readable and hence its meaning is more readily understood. But, since there is a

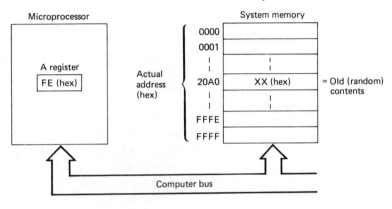

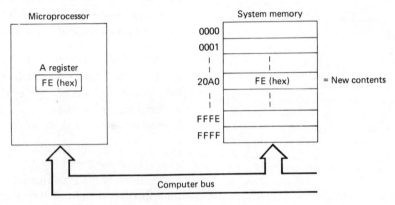

Figure 3.1 Execution of instruction LD (20A0),A

one-for-one correspondence between the symbolic notation form and the binary-coded form, it is also very straightforward to convert from one form to the other. All the instructions presented in this and subsequent chapters will be written in their symbolic notation form, but the conversion process between the two forms will also be described.

Q3.1

3.4 Classification of instructions

A typical microprocessor may execute perhaps a hundred or more different machine instructions and these are collectively referred to as the *instruction set* of that device. To understand the meaning of each instruction in the complete set clearly seems a daunting task but fortunately each instruction can be classified as being a member of just one of four different groups:

1 Data movement.
2 Data manipulation.

3   Transfer of control.
4   Input-output.

Instructions in the *data movement* group move or transfer data either between the various registers within the microprocessor or between a microprocessor register and a memory location. For example,

    LD   A,B

results in the current contents of register B being *moved* or *loaded* into register A. Notice that the destination address (that is, the accumulator or register A in this example) is always specified *before* the source address (register B). Data movement instructions are discussed later in this chapter.

Instructions in the *data manipulation* group perform arithmetic and logic operations on data which are either in a specified register or a memory location. For example,

    ADD   A,B

results in the register A containing the sum of its current contents and the current contents of register B. Notice how register A accumulates the result. This is true of most data manipulation instructions. The latter are discussed in detail in Chapter 4.

Instructions in the *transfer of control* group normally result in the microprocessor breaking the sequential order of execution of the instructions which comprise a program and instead cause the microprocessor to branch to a different part of the program to obtain the next instruction to be executed. Thus

    JP   20FF

results in the microprocessor branching (jumping) unconditionally out of sequence to memory address 20FF (hex) to fetch the next instruction to be executed. As will be seen, transfer of control instructions result in a high degree of flexibility when writing a program for a microprocessor. They are described in Chapter 5.

Finally, instructions in the *input/output* group move data between the various I/O 'ports' of the system and a microprocessor register – usually the accumulator or register A. For example,

    OUT   (20),A

results in the contents of register A being transferred to the output port with the hexadecimal address 20. The I/O instructions are discussed in Chapter 6.

The actual instruction mnemonics selected for example purposes in this and subsequent chapters are those available with the Zilog Z80 microprocessor. This is a widely used device and both its architecture

Q3.2
and type of instructions are typical of those available with most 8-bit microprocessors.

3.5 Instruction addressing modes

The source and destination addresses used with an instruction clearly vary. Data movement instructions, for example, may specify a microprocessor register or a memory location as its source or destination address. Alternatively, as will be seen, a data movement instruction may specify an actual data value in place of an address.

The type of source and destination addresses utilised by an instruction is determined by the *instruction addressing mode* and all microprocessors provide a variety of differing address modes. The range of addressing modes provided by a particular microprocessor is important since it can often result in considerable flexibility when writing a program and also leads to programs that require fewer instructions to implement a given task.

The four main types of addressing mode used in microprocessor systems are:

1 Register addressing.
2 Immediate addressing.
3 Direct addressing.
4 Register indirect addressing.

The first two are used primarily for data movement and data manipulation instructions which involve only internal microprocessor registers. The second two are used primarily for data movement and data manipulation instructions which involve a microprocessor register and a location in memory.

Q3.3
To illustrate the use of each of the above addressing modes, typical machine instructions from the data movement group of instructions are now considered.

3.6 Data movement instructions

Instructions in the data movement group provide the facility for moving data from one register or memory location in the microcomputer to another. Hence instructions are provided both to move data between the various registers within the microprocessor and also to move data between a microprocessor register and a location within the memory.

Since many of the instructions in the data movement group operate on or affect the state of the various registers within the microprocessor, the main registers of the Z80 are shown as an example in

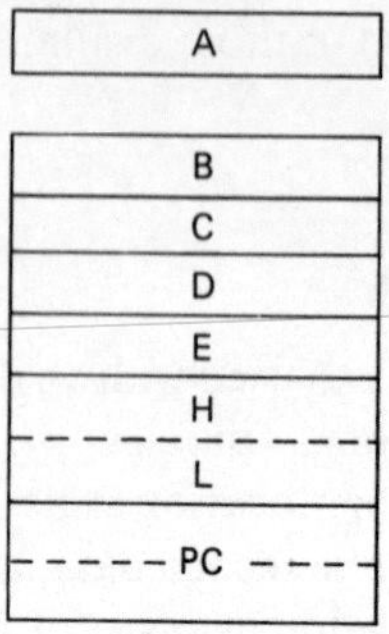

Figure 3.2 Main registers in the Zilog Z80 microprocessor:

A   is the 8-bit arithmetic or accumulator register

B,C,D,E   are four 8-bit general-purpose registers

H,L   are two 8-bit registers which are normally used to form a 16-bit memory address register

PC   is the program counter register which contains a 16-bit memory address which points to the next instruction to be executed

Figure 3.2. The use of each register will be described as the various machine instructions are introduced.

Register addressing

The instructions which use the register addressing mode are those used to move data between the internal microprocessor registers and hence the instruction source and destination addresses both specify which of these registers are involved in the transfer.

For example,

```
LD   B,A
```

↑ Operation (load)   ↖ Operand (B = destination; A = source)

This results in the contents of register A being moved or loaded into register B. The contents of register A remain unchanged. This is often written as:

```
(B) ← (A); (A) unchanged
```

where the brackets are used to indicate 'contents of'.

Alternatively, the action of an instruction may be illustrated by means of an entry in a *trace table*. Each entry in a trace table contains, in addition to the instruction, a list of the contents of the various registers before and after execution of the instruction. This form of representation is therefore particularly helpful for observing the effect of the actions associated with those instructions which involve internal registers. Thus a trace table entry for the above instruction is as shown in Figure 3.3(a).

For clarity, those registers which are not affected by this instruction are shown as xx in the table to indicate 'don't care' states. It is assumed that the contents of register B prior to execution of the instruction are E7 (hex) and hence, after execution, register A also contains E7 (hex) and the contents of register B are unchanged.

Symbolic instruction	Microprocessor register contents															
	Before execution								After execution							
	A	B	C	D	E	H	L	PC	A	B	C	D	E	H	L	PC
LD B,A	xx	E7	xx	xx	xx	xx	xx	xxxx	E7	E7	xx	xx	xx	xx	xx	xxxx

(a) LD B,A

	A	B	C	D	E	H	L	PC	A	B	C	D	E	H	L	PC
LD C,B	xx	E7	xx	xx	xx	xx	xx	xxxx	xx	E7	E7	xx	xx	xx	xx	xxxx

(b) LD C,B

	A	B	C	D	E	H	L	PC	A	B	C	D	E	H	L	PC
EX DE,HL	xx	xx	xx	20	80	28	00	xxxx	xx	xx	xx	28	00	20	80	xxxx

(c) EX DE,HL

	A	B	C	D	E	H	L	PC	A	B	C	D	E	H	L	PC
LD A,7F	xx	xx	xx	xx	xx	xx	xx	xxxx	7F	xx	xx	xx	xx	xx	xx	xxxx

(d) LD A,7F

	A	B	C	D	E	H	L	PC	A	B	C	D	E	H	L	PC
LD HL,2800	xx	xx	xx	xx	xx	xx	xx	xxxx	xx	xx	xx	xx	xx	28	00	xxxx

(e) LD HL,2800

	A	B	C	D	E	H	L	PC	A	B	C	D	E	H	L	PC
LD DE,2080	xx	xx	xx	xx	xx	xx	xx	xxxx	xx	xx	xx	20	80	xx	xx	xxxx

(f) LD DE,2080

Figure 3.3 A trace table showing some different data movement instructions

Another example,

```
LD  C,B
```

means load the contents of register B into register C:

```
(C) ← (B); (B) unchanged
```

and hence a typical trace table entry for this instruction is as shown in Figure 3.3(b). Again, the contents of register B prior to execution of this instruction are assumed to be E7 (hex).

In addition, there are a limited number of data movement instructions which involve combined 16-bit register pairs, e.g. the D,E and H,L register pairs with the Z80. A typical instruction is

```
EX  DE,HL
```

which results in the contents of the D,E register pair being exchanged with the contents of the H,L register pair. This is represented as

```
(D)(E) ↔ (H)(L)
```

and hence a trace table entry for this instruction might be as shown in Figure 3.3(c). The figure assumes the initial contents of the D,E and H,L register pairs are 2080 (hex) and 2800 (hex), respectively. Thus, after execution of the instruction, register pair D,E will contain 2800 (hex) and H,L will contain 2080 (hex).

Immediate addressing

With this type of addressing mode, the operand part of the instruction is not used to specify a microprocessor register or memory
location where the source data is located but instead it specifies an
actual data value. Thus the source *data* associated with the operation
part of the instruction is contained *within the instruction itself* and is
therefore said to be *immediately* available. The data value is normally
specified in its hexadecimal form. For example:

LD   A,7F

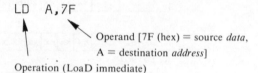

Operand [7F (hex) = source *data*,
A = destination *address*]

Operation (LoaD immediate)

This results in the data value 7F (hex) in this example being loaded
into register A:

(A) ← 7F(hex) or A ← 01111111 (binary)

An example trace table entry for this instruction is therefore as shown
in Figure 3.3(d). Notice that the contents of register A prior to
execution of the instruction are shown as 'don't care' since,
irrespective of its initial contents, it will always be 7F (hex) after
execution.

Similarly, a 16-bit register pair (BC, DE or HL for the Z80) may be
specified as the destination address and consequently these instructions require two bytes of immediate data. For example:

LD   HL,2800

results in the 16-bit H,L register pair being loaded with immediate
data 2800 (hex):

(H) ← 28 (hex)
(L) ← 00 (hex)
(H)(L) ← 2800 (hex)

Similarly

LD   DE,2080; (D)(E) ← 2080 (hex)

and hence the trace table entries in Figure 3.3(e) and 3.3(f),
respectively.

Program example 3.1: register data transfer

The program example shown in Figure 3.4 uses a combination of the above
instructions firstly to load a value into register A using immediate addressing
and then to load this value into two further registers, B and C, using register
addressing. Register pairs H,L and D,E are then loaded with 16-bit data
values using immediate addressing. Finally, their contents are exchanged

(a) Flowchart

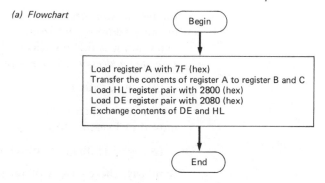

(b) Program code

Symbolic instructions			Action
Mnemonic	OP 1	OP 2	
LD	A	7F	(A) ← 7F (hex)
LD	B	A	(B) ← 7F (hex)
LD	C	B	(C) ← 7F (hex)
LD	HL	2800	(H)(L) ← 2800 (hex)
LD	DE	2080	(D)(E) ← 2080 (hex)
EX	DE	HL	(D)(E) ↔ (H)(L)

(c) Trace table

Symbolic instructions	Microprocessor register contents															
	Before execution								After execution							
	A	B	C	D	E	H	L	PC	A	B	C	D	E	H	L	PC
LD A,7F	xx	xx	xx	xx	xx	xx	xx	xxxx	7F	xx	xx	xx	xx	xx	xx	xxxx
LD B,A	7F	xx	xx	xx	xx	xx	xx	xxxx	7F	7F	xx	xx	xx	xx	xx	xxxx
LD C,B	7F	7F	xx	xx	xx	xx	xx	xxxx	7F	7F	7F	xx	xx	xx	xx	xxxx
LD HL,2800	7F	7F	7F	xx	xx	xx	xx	xxxx	7F	7F	7F	xx	xx	28	00	xxxx
LD DE,2080	7F	7F	7F	xx	xx	28	00	xxxx	7F	7F	7F	20	80	28	00	xxxx
EX DE,HL	7F	7F	7F	20	80	28	00	xxxx	7F	7F	7F	28	00	20	80	xxxx

Figure 3.4 Program example 3.1

using register addressing. Part *(a)* of the figure shows a *flow diagram or flowchart* for the program and parts *(b)* and *(c)* two alternative ways of representing the list of program instructions and their actions. A full description of the use of flowcharts to illustrate the flow or sequence of execution of a program will be given in Chapter 5 when the transfer of control instructions are discussed. One will be included with each of the program examples which follow, however, since, as can be seen, the descriptive information contained within a flowchart forms a convenient way of illustrating the actions of a program.

Q3.4, 3.5

Direct (extended addressing)

Direct addressing is used when the data value associated with the instruction is located in the system memory. Thus the operand part of the instruction specifies the address of a memory location. Since all memory addresses are 16 bits long, however, the address requires two bytes and it is for this reason that this mode is often referred to as *extended addressing*. For example:

```
LD   A,(2800)
```

results in the current contents of memory location 2800 (hex) being moved or loaded into register A. It is therefore expressed as

```
(A) ← (2800); (2800) unchanged
```

Similarly, the contents of register A may be stored in a specified memory location. For example:

```
LD   (2800),A;  (2800) ← (A);  (A) unchanged
```

results in the contents of register A being stored in memory location 2800 (hex). The contents of register A remain unchanged.

As the next section will show, the H,L register pair is frequently used as a combined 16-bit memory address register and consequently two instructions are provided to enable the H,L register pair to be loaded using a single instruction and direct addressing, e.g.

```
LD   HL,(2800)
```

This results in register L being loaded with the contents of memory location 2800 and register H being loaded with the contents of the next consecutive memory location, i.e. 2801 in this example. This is therefore expressed as:

```
(L) ← (2800)
(H) ← (2801)
```

Similarly, the contents of H and L can be stored in two consecutive memory locations using:

```
LD   (2802),HL   (2802) ← (L);  (L) unchanged
                 (2803) ← (H);  (H) unchanged
```

Program example 3.2: direct addressing

The program of Figure 3.5 uses a combination of immediate and direct addressing first to store a data value into two consecutive memory locations, then to load register pair H,L with this data. Finally, the same data are stored into a pair of different memory locations using direct addressing.

Register indirect addressing

Register indirect addressing is an alternative to direct addressing for writing (storing) and reading (loading) data to and from a memory

(a) Flowchart

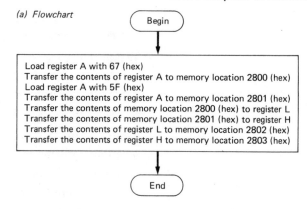

```
Load register A with 67 (hex)
Transfer the contents of register A to memory location 2800 (hex)
Load register A with 5F (hex)
Transfer the contents of register A to memory location 2801 (hex)
Transfer the contents of memory location 2800 (hex) to register L
Transfer the contents of memory location 2801 (hex) to register H
Transfer the contents of register L to memory location 2802 (hex)
Transfer the contents of register H to memory location 2803 (hex)
```

(b) Program code

Symbolic instructions	Action
LD A,67	(A) ← 67 (hex)
LD (2800),A	(2800) ← 67 (hex)
LD A,5F	(A) ← 5F (hex)
LD (2801),A	(2801) ← 5F (hex)
LD HL,(2800)	(L) ← (2800), i.e. 67 (hex)
	(H) ← (2801), i.e. 5F (hex)
LD (2802),HL	(2802) ← 67 (hex)
	(2803) ← 5F (hex)

Figure 3.5   Program example 3.2

location. As will be seen in subsequent chapters, this provides a far more convenient and efficient method for accessing a list of values representing, say, a table in memory. Consequently, since this is a common requirement in a number of microcomputer applications, the use of indirect addressing can often result in substantial savings in the amount of memory required for the program.

Using register indirect addressing the data value is either read from or written to the memory location, the address of which is the current 16-bit contents of register pair H,L. The instruction does not therefore contain the memory address but instead it is *implied* in the operation code that the address to be used is the current contents of the register pair H,L. The memory address is therefore obtained *indirectly*. For example:

```
LD  A,(HL)
```

results in register A being loaded with the contents of the memory location whose address is the current contents of register pair H,L. This is represented as

```
(A) ← ((H)(L))
```

Similarly,

```
LD  (HL),B; ((H)(L)) ← (B); (B) unchanged
```

results in the contents of B being moved to the memory location currently specified in register pair H,L.

In addition, there is also an instruction to enable an immediate data value to be stored directly into a memory location. Again the memory address is stored in the register pair H,L. Thus

```
LD  (HL),AA;  ((H)(L)) ← AA (hex)
```

results in the data value AA (hex) being stored in the memory location whose address is in register pair H,L.

Program example 3.3: register indirect addressing

The program of Figure 3.6 uses a combination of immediate and register indirect addressing. The memory address is first loaded into register pair H,L using immediate addressing and then a value is stored in this memory location using register indirect addressing. Finally, the value is loaded into two further registers, again using register indirect addressing.

Q3.6, 3.7

(a) Flowchart

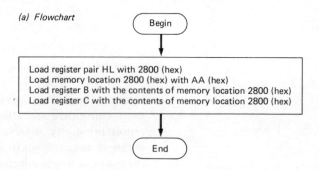

Begin

Load register pair HL with 2800 (hex)
Load memory location 2800 (hex) with AA (hex)
Load register B with the contents of memory location 2800 (hex)
Load register C with the contents of memory location 2800 (hex)

End

(b) Program code

Symbolic instructions	Action
LD HL,2800	(L) ← 00 (hex)
	(H) ← 28 (hex)
LD (HL),AA	(2800) ← AA (hex)
LD B,(HL)	(B) ← (2800), i.e. (B) ← AA (hex)
LD C,(HL)	(C) ← (2800), i.e. (C) ← AA (hex)

Figure 3.6   Program example 3.3

3.7 The Zilog Z80 instruction set

The instructions so far introduced are by no means the only instructions available in the data movement group. Indeed, they are intended only as examples from the complete list of available instructions. For instance, in the data movement group there are normally instructions to enable data to be moved between any pair of registers (A, B, C, D, E, H and L for the Z80). Also, immediate data may be loaded into each register.

The aim of this and subsequent chapters is simply to introduce examples of instructions from each instruction group, together with their meaning and application areas, so that the reader is able firstly to understand the function of each instruction group and secondly to select those instructions from the group required to perform a specific task. A more complete list of instructions for the Z80 is given as an example in Appendix 1.

3.8 The translation process

A microprocessor executes instructions stored in a binary-coded form in its program memory. Consequently, before any program can be executed, including the examples above, they must first be converted from their symbolic notation form into their equivalent binary-coded form. All microcomputer systems, therefore, whether large development systems or small single-board prototyping systems, normally incorporate a suite of programs (system programs) which have been designed to translate programs written in symbolic notation form (source code) into their equivalent executable binary (object) code.

In many small microcomputer prototyping systems, however, the resident operating program (normally called the monitor) cannot perform a complete translation process of the source code and instead each instruction must first be converted by the programmer into an intermediate hexadecimal form. The monitor then translates each pair of hexadecimal characters into the corresponding 8-bit binary pattern. This process is illustrated in Figure 3.7.

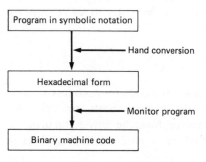

Figure 3.7 The translation process

The translation process from symbolic notation to hexadecimal form clearly has to be frequently performed by the programmer. Consequently the manufacturer's list of symbolic machine instructions normally contains the corresponding hexadecimal code for each instruction. For example, the list of Z80 instructions shown in Appendix 1 also includes the hexadecimal equivalent of each instruction.

Hand conversion

It can readily be deduced from Appendix 1 that each instruction requires from one to three bytes to be represented in hexadecimal form. For example, the instruction

```
LD   B,A; (B) ← (A)
```

requires only a single byte:

| 4 | 7 | Operation |

The instruction

```
LD   A,7F; (A) ← 7F (hex)
```

requires two bytes:

3	E	Operation
7	F	Immediate data, e.g. 7F (hex)

and the instruction

```
LD   HL,2800; (H)(L) ← 2800 (hex)
```

requires three bytes:

2	1	Operation
0	0	Least significant byte = (L)
2	8	Most significant byte = (H)

Notice that when the operand part of an instruction specifies a 16-bit (2-byte) value, the least significant byte (00 in the above example) is always specified first.

Coding sheets

In addition to providing the hexadecimal code for each instruction, most manufacturers also provide coding sheets to help the programmer to write a program and perform the conversion process. The layout of a typical manufacturer's program coding sheet is shown in Figure 3.8.

The program illustrated is Program example 3.1. As can be seen, the symbolic notation form of the program and the associated comments are listed as before. The LABEL field associated with each instruction is used for branch instructions and hence it is left blank in this example. Its use will be discussed in a later chapter.

It has been assumed that the program will eventually be stored in memory starting at address 2000 (hex) and hence, as each instruction is converted into its hexadecimal form, it is assigned the next sequential memory address starting at this value. The hexadecimal form of each instruction can be obtained from Appendix I.

3.9   Loading and running a program

Although for most microprocessor applications the program in the final system will be permanently stored in, say, read-only memory, during the development and testing phase it is helpful to be able to **Q3.8** readily load, run and, if necessary, modify the program.

Memory		Symbolic Instruction				Comments
Address	Contents	Label	Mnemonic	OP 1	OP 2	
2000	3E					
2001	7F		LD	A	7F	(A) ← 7F (hex)
2002	47					
			LD	B	A	(B) ← (A)
2003	4F					
			LD	C	B	(C) ← (B)
2004	21					
2005	00		LD	HL	2800	(H)(L) ← 2800 (hex)
2006	28					
2007	11					
2008	80		LD	DE	2080	(D)(E) ← 2080 (hex)
2009	20					
200A	EB					
			EX	DE	HL	(D)(E) ↔ (H)(L)

Figure 3.8 Manufacturer's program coding sheet

Consequently manufacturers normally offer complete made-up single-board microcomputer systems which, in addition to the basic microcomputer, incorporate facilities for loading, running and testing small programs that have been translated into hexadecimal form.

Essentially, a single-board system contains the microprocessor, a quantity of read-only memory (ROM) for holding the system monitor program and a quantity of random-access memory (RAM) for storing both the program to be tested and any data values associated with the program. A hexadecimal keypad is then provided on the front of the board to enter the program and an associated hexadecimal display to display the contents of selected memory locations and registers during program testing. Finally, additional keys for entering commands to the monitor program are provided and a description of the commands normally available and their uses is now given.

Monitor commands

The layout of a typical keypad and an associated hexadecimal display for use with a single-board system is shown in Figure 3.9. As can be

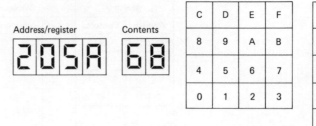

Figure 3.9   Keypad and display of a single board system

seen, in addition to the sixteen hexadecimal keys, the keypad also contains five monitor command keys.

Commands to the monitor program are entered by pressing the selected command key and the response of the monitor is displayed on the hexadecimal display. All the displayed digits are in hexadecimal code and are split into two parts: the address/register field and the data or contents field. Thus as an example, the figure illustrates a typical memory address 205A (hex) and the corresponding contents 68 (hex).

The five monitor command keys shown in the figure are typical of those available with a single-board system. They are: reset; substitute memory; run; examine register; single step (or trace).

Reset   This command enables the user to force the monitor program to restart execution from its beginning and, after giving the command, the monitor is ready to accept further commands. It is normally used when power is first applied or when the user wishes to regain control of the system.

Substitute memory   This command allows the user to examine the contents of successive memory locations and, if required, to modify their contents. This command is particularly useful therefore for entering a user's program into RAM. The user first writes the program in symbolic notation, converts this into hexadecimal form with the aid of a coding sheet and then uses the 'Substitute Memory' command to load the resulting hexadecimal code into RAM. The 'Substitute Memory' command also allows the user to modify specific RAM locations – and hence program instructions – if any errors are found in the program during execution.

Run   After a program has been written, coded and loaded into memory, it is executed or run using the 'Run' command. To run a program that has already been stored in RAM, the 'Run' key is pressed followed by the 4-digit address in memory of the first instruction in the program. This is entered using the hexadecimal keypad.

Once the program is running, the monitor only regains control of the system if either the 'Reset' key is pressed or an instruction in the program is executed to force the microprocessor to branch to the start of the monitor program.

Examine register This command allows the user to display and, if required, modify the contents of each of the registers. This is a particularly useful facility when used with the 'Single step' command since it enables the user to monitor the status of the microprocessor during user program execution and hence can be used to identify possible program errors.

When a program has been loaded into RAM it is usually executed using the 'Run' command as outlined above. If the program contains errors, however, and does not perform the required task, it is necessary to find the erroneous instructions so that they may be corrected. This can often be a difficult and, without any system aids, time-consuming task.

Single step or trace This command is such a system aid since it enables a user to examine the state of the complete system – registers, memory locations, etc. – as each program instruction is executed. Thus program instructions that do not produce the required effect may be readily identified.

To step through a program which is already stored in memory, the 'Single step' key is first pressed, followed by the address in memory of the first instruction. This causes the processor to execute the first instruction but, instead of continuing execution, the program is suspended and the monitor program is again run.

Since control is now with the monitor, before proceeding the user may, if required, examine the contents of selected registers and/or memory locations to verify correct operation of the previous instruction. Assuming this is correct, the user may then execute the next instruction by pressing the 'Single step' key as before.

Questions

3.1 Explain the meaning of the following terms:
   (a) Symbolic notation.
   (b) Operation mnemonic.
   (c) Symbolic address.

3.2 With the aid of examples describe the actions performed by an instruction from each of the following instruction groups:
   (a) Data transfer.
   (b) Data manipulation.
   (c) Transfer of control.
   (d) Input/output.

3.3 With the aid of examples describe the meaning of the following addressing modes used in microcomputer systems:
(a) Register addressing.
(b) Immediate addressing.
(c) Direct addressing.
(d) Register-indirect addressing.

3.4 Illustrate the actions of the following program instructions by means of a trace table and hence derive the contents of registers A, B, C, D, E, H and L after the program has run:

```
LD   DE,2800
LD   A,D6
LD   B,D
LD   C,E
LD   HL,28FF
EX   DE,HL
```

3.5 Write a program to:
(a) Load the immediate data value FF (hex) into the register A.
(b) Load the immediate data value 86 (hex) into the register B.
(c) Load the contents of register A into register D.
(d) Load the contents of register B into register E.
(e) Load the immediate data value 28A6 (hex) into register pair H,L.
(f) Exchange the contents of register pair D,E and register pair H,L.

3.6 Write a program to:
(a) Load the immediate data value AA (hex) into register A.
(b) Store the contents of register A in memory location 2800 (hex) using direct addressing.
(c) Load the immediate data value 00 (hex) into register A.
(d) Load register pair H,L with immediate data 2800 (hex).
(e) Load register A with the contents of memory location 2800 (hex) using indirect addressing.

3.7 Derive the contents of memory locations 2800 (hex) and 2801 (hex) after the following program has run:

```
LD   B,EE
LD   C,FF
LD   HL,2801
LD   A,B
LD   (2800),A
LD   (HL),C
```

3.8 List the above programs on a coding sheet similar to the one shown in Figure 3.7 and derive the hexadecimal code for each program using the hexadecimal codes given in Appendix I.

Chapter 4 Arithmetic and logic instructions

Objectives of this chapter *When you have completed studying this chapter you should be able to:*

1   *Understand the two's complement form of number representation and be able to perform addition and subtraction operations on numbers represented in this form.*
2   *Know the different types of addition and subtraction instructions provided with a microprocessor and be able to write and understand programs using these instructions.*
3   *Know the various flag bits used in a flags register and be able to describe an application of the carry and the auxiliary carry flags.*
4   *Understand the binary-coded decimal form of number representation and be able to perform addition and subtraction operations on numbers represented in this form.*
5   *Know the function of the decimal adjust instruction and be able to write and understand a program using this instruction.*
6   *Know the logical instructions AND, OR and XOR provided with a microprocessor and be able to explain an application of each type of instruction.*
7   *Know the actions performed by the shift and compare instructions.*

4.1   Introduction

Microprocessors may be used in a variety of different applications. Consequently, depending on the application, the binary data stored within the microcomputer may represent a variety of different parameters. In one application, for example, the 8-bit contents of a register or memory location may represent a signed value or number (+62, –27, etc.) whilst in another application it may indicate the state of, say, eight controlled two-state devices (logical 1 = on, 0 = off).

Hence, when manipulating this data, a typical operation on a signed value might be to add or subtract another similar value to it whilst if the stored value indicates the state of, say, eight controlled devices, a typical operation might be to test if a specific bit (and hence device) is a logical 1 (on) or 0 (off).

Instructions in the data manipulation group, therefore, not only include the normal arithmetic instructions add, subtract, etc., but also the logical instructions AND, OR, etc. These, as will be seen, allow a programmer to test and manipulate individual binary digits within a collection of, say, 8 bits. The logical instructions are, therefore, particularly useful in applications involving the manipulation of individual binary bits.

4.2   Signed number representation

Before considering the different arithmetic instructions provided with a microprocessor it is first necessary to understand how signed numbers are stored within a microcomputer. The standard method used is the *two's complement notation* since, as will be seen, when performing arithmetic operations on data stored using this method the correct signal result is automatically produced. This results in considerably simpler electronic circuitry to implement the arithmetic logic unit (ALU).

Using the two's complement notation, the most significant bit, S, of each binary number is used to indicate the sign of the number. Thus with an 8-bit system each number is represented as follows:

m.s.			l.s.
S	6	- - - - - - - - -	0

S = 0 for positive numbers and zero
S = 1 for negative numbers

For positive numbers, the remaining seven digits then indicate the magnitude of the number. For example:

S	2^6	2^5	2^4	2^3	2^2	2^1	2^0	= weighting (power of 2)
0	0	0	1	0	1	0	1	= +21 (decimal)
0	1	0	0	1	0	1	0	= +74
0	1	1	0	0	0	0	1	= +97

For negative numbers, the remaining seven digits represent the magnitude of the number in its two's complement form. To obtain the two's complement of a number, the complete binary number – including the sign bit – is first inverted (complemented) and the resulting binary value incremented by 1. For example:

$+21 = 0\ 0010101$ ———→

invert

$1\ 1101010$ ←———

increment (add 1)

Therefore $-21 = 1\ 1101011$ ———→

Table 4.1   *Range of 8-bit two's complement numbers*

Decimal number	Two's complement form
+127	0 1111111
+126	0 1111110
.	.
.	.
.	.
+ 3	0 0000011
+ 2	0 0000010
+ 1	0 0000001
0	0 0000000
- 1	1 1111111
- 2	1 1111110
- 3	1 1111101
.	.
.	.
.	.
-127	1 0000001
-128	1 0000000

Similarly:

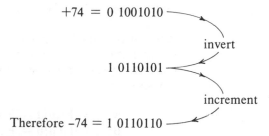

Therefore $-74 = 1\ 0110110$

Note that after the complement process the most significant bit (the sign bit) is automatically a 1. Hence for an 8-bit system, the range of possible numbers is as shown in Table 4.1. Note that the maximum positive 8-bit number is $+127$ and the maximum negative number is -128.

Q4.1, 4.2

4.3 Arithmetic instructions

An electronic calculator is always concerned with performing arithmetic operations on signed numbers. Consequently, in addition to addition and subtraction operations, most calculators also provide multiplication, division and perhaps some more sophisticated arithmetic functions. Since in many microprocessor applications the binary values being manipulated represent non-numerical information, however, the majority of microprocessors only provide the basic arithmetic instructions of addition and subtraction. This means that with most 8-bit microprocessors multiplication and division cannot be performed using a single instruction and instead the programmer must use a group or sequence of instructions to perform these operations. The basic arithmetic instructions of addition and subtraction are now considered.

The addition operation

The addition of two binary digits (bits) is summarised in Table 4.2. A figure that tabulates all the possible combinations of the inputs – Bit 1 and Bit 2 in this example – and lists the corresponding outputs – sum and carry – is referred to as a *truth table*.

It can readily be deduced from the table that, since the sum bit can only be a (logical) 0 or 1, when summing two one's together the sum bit is 0 and a *carry bit* is generated. The carry bit must then be added to the next higher order pair of bits. Hence when performing the addition operation, it is necessary to add the possible carry bit (the carry-in) that may have been generated from the addition of the previous pair of bits.

Table 4.2 *Addition of 2 binary bits*

Bit 1	Bit 2	Sum	Carry
0	0	0	0
0	1	1	0
1	0	1	0
1	1	0	1

The complete addition process is therefore as summarised in the truth table shown in Table 4.3.

It is now possible to apply the information contained in Table 4.3 to the addition of two full binary numbers. The following example illustrates the addition of two positive numbers, A and B, to produce a positive sum:

$$A = 0\ 0101011 = +\ 43\ \text{(dec)}$$
$$B = \underline{0\ 0110010} = +\ 50\ \text{(dec)}$$
$$\text{carry-in} = \overline{0\ 1000100}$$

carry-out (Cout) ———— carry-in (Cin)

$$\text{carry-out} = 0\ 0100010$$

$$A + B \quad = \underline{0\ 1011101} = +\ 93\ \text{(dec)}$$

As was mentioned earlier, providing negative numbers are represented in their two's complement form, the same addition operation may be performed and the correct two's complement signed result will automatically be obtained. This is perhaps best illustrated by means of examples. The first example illustrates the addition of a negative and a positive number to produce a negative result; the second, the addition of a negative and a positive number to produce a positive result and the third, the addition of two negative numbers to produce a negative result.

$$A = 0\ 0101011 = +\ 43$$
$$B = \underline{1\ 1001110} = -\ 50$$

$$\text{carry-in} = \overline{0\ 0011100}$$

Cout ———— Cin

$$\text{carry-out} = 0\ 0001110$$

$$A + B = \underline{1\ 1111001} = -\ 7$$

$$A = 1\ 1010101 = -\ 43$$
$$B = \underline{0\ 0110010} = +\ 50$$

$$\text{carry-in} = \overline{1\ 1100000}$$

Cout ———— Cin

$$\text{carry-out} = 1\ 1110000$$

$$A + B = \underline{0\ 0000111} = +\ 7$$

Table 4.3 *Addition of 2 bits and a carry-in bit*

Bit 1	Bit 2	Carry-in	Sum	Carry-out
0	0	0	0	0
0	0	1	1	0
0	1	0	1	0
0	1	1	0	1
1	0	0	1	0
1	0	1	0	1
1	1	0	0	1
1	1	1	1	1

$$A = 1\ 1010101 = -43$$
$$B = 1\ 1001110 = -50$$

$$\text{carry-in} = 1\ 0111000$$

Cout ⟍⟍⟍⟍⟍⟍⟍ Cin

$$\text{carry-out} = 1\ 1011100$$

$$A + B = 1\ 0100011 = -93$$

The decimal equivalents of the negative results can readily be confirmed by performing the complement process on the binary results.

Add instructions

Normally, the addition instructions provided with a microprocessor involve the contents of register A and a second value specified using either register, immediate or register indirect addressing. Some examples from the Z80 instruction set which use each of these addressing modes are now given.

Register addressing An example of an instruction which uses register addressing is

```
ADD   A,B
```

This results in the current contents of register B being added to the current contents of register A. The result replaces the current contents of register A and the contents of register B remain unchanged. This is represented as:

```
(A) ← (A) + (B); (B) unchanged
```

Immediate addressing An example which uses immediate addressing is

```
ADD   A,8E
```

This results in the immediate data value contained within the instruction, 8E (hex) in this example, being added to the current contents of register A, i.e.

```
(A) ← (A) + 8E (hex)
```

Register indirect addressing   An example which uses register indirect addressing is

```
ADD   A,(HL)
```

This results in the contents of the memory location whose address is specified in register pair HL being added to the current contents of the A register. The contents of the memory location remain unchanged:

Q4.3(a)   `(A) ← (A) + ((H)(L)); ((H)(L)) unchanged`

The subtraction operation

The subtraction of two binary digits is perhaps best explained by considering the subtraction of two numbers in the decimal system. Consider the following example:

$$
\begin{array}{rr}
A = & 465 \\
B = & -173 \\
Borrow = & 100 \\
\hline
A - B = & 292 \\
\end{array}
$$

When the second digit is less than or equal to the first digit the subtraction can be performed straight away and no borrow is required (for example, 5 minus 3). When the second digit is greater than the first, however, (6 minus 7 in the example) it is necessary to borrow a digit from the next higher order position. This is therefore called a *borrow-in* and in the decimal system the borrow is equal to ten times the magnitude of the other digits in that column. The subtraction operation in that column then becomes $(10 + 6) - 7$ which is equal to a difference of 9 with a borrow to be carried over to the next column. This is therefore called a *borrow-out* and, in turn, must be added to the second digit (1 in the example) before the latter is subtracted from the first digit (4 in the example).

To summarise, when performing the subtraction operation of two digits it is necessary to consider a possible borrow-in from a previous

Table 4.4 *Subtraction of 2 bits and a borrow-in bit*

Bit 1	Bit 2	Borrow-in	Difference	Borrow-out
0	0	0	0	0
0	0	1	1	1
0	1	0	1	1
0	1	1	0	1
1	0	0	1	0
1	0	1	0	0
1	1	0	0	0
1	1	1	1	1

column. Also, if the sum of the second digit and the borrow-in is greater than the first digit then a borrow-out is generated.

The subtraction of two binary digits and a possible borrow-in is therefore as shown in the truth table of Table 4.4. Bit 2 and the borrow-in are first added together and the sum is then subtracted from Bit 1. If the sum is greater than Bit 1 then a borrow-out is required. This is equivalent to two bits in the binary system and is in turn added to Bit 1 before the subtraction operation.

It is now possible to apply the information contained in Table 4.4 to the subtraction of two binary numbers. This can be illustrated as follows:

$$A = 0\ 0110010 = +50\ \text{(dec)}$$
$$B = 0\ 0101011 = +43\ \text{(dec)}$$

$$\text{Borrow-in} = 0\ 0011110$$

Borrow-out (Bout) ⟵ ⫿⫿⫿⫿⫿⫿⫿ ⟶ Borrow-in (Bin)

$$\text{Borrow-out} = 0\ 0001111$$

$$A - B = 0\ 0000111 = +7\ \text{(dec)}$$

The first number in the above example, A, was specifically selected to be greater than the second number, B, and consequently, since both numbers were positive, the difference was also positive. As was mentioned earlier, however, providing the two numbers are represented in their two's complement form the same bit-for-bit subtraction operation can be performed on the two numbers and the correct two's complement signed result will be obtained. This is illustrated in the following examples:

$$A = 0\ 0101011 = +\ 43$$
$$B = 1\ 1001110 = -\ 50$$

Borrow-in $= 1\ 0111000$

Bout ⟵ ⟍⟍⟍⟍⟍⟍⟍ ⟶ Bin

Borrow-out $= 1\ 1011100$

$$A - B = 0\ 1011101 = +\ 93\ [+43-(-50)]$$

$$A = 1\ 1010101 = -\ 43$$
$$B = 0\ 0110010 = +\ 50$$

Borrow-in $= 0\ 1000100$

Bout ⟵ ⟍⟍⟍⟍⟍⟍⟍ ⟶ Bin

Borrow-out $= 0\ 0100010$

$$A - B = 1\ 0100011 = -\ 93\ [-43-(+50)]$$

$$A = 1\ 1010101 = -\ 43$$
$$B = 1\ 1001110 = -\ 50$$

Borrow-in $= 0\ 0011100$

Bout ⟵ ⟍⟍⟍⟍⟍⟍⟍ ⟶ Bin

Borrow-out $= 0\ 0001110$

$$= 0\ 0000111 = +\ 7\ [-43-(-50)]$$

Subtract instructions

The subtract instructions normally available with a microprocessor are similar to the add instructions described previously. They involve the contents of register A and a second value specified using either register, immediate or register indirect addressing. The microprocessor simply performs the equivalent of the bit-by-bit subtraction as outlined above. Some examples from the Z80 instruction set are as follows:

Register addressing:

```
SUB   B
```

That is:

```
(A) ← (A) - (B); (B) unchanged
```

Immediate addressing:

    SUB   8E

That is:

    (A) ← (A) - 8E (hex)

Register indirect addressing:

    SUB   (HL)

That is:

    (A) ← (A) - ((H)(L)); ((H)(L)) unchanged

Increment and decrement

A very common operation in many application programs involving microprocessors is to increment or decrement the current contents of a microprocessor register or memory location by unity. Consequently, although it is possible to perform these operations using immediate addressing, most microprocessors also include instructions to perform these functions. For example,

    INC   A

results in the current contents of register A being incremented by unity, i.e.

    (A) ← (A) + 1

Similarly,

    DEC   B

results in the contents of register B being decremented by unity:

    (B) ← (B) - 1

A particularly useful instruction when accessing and manipulating a list of values stored in memory is the instruction

    INC   HL

This results in the combined contents of register pair H,L (the memory address register) being incremented by unity:

    (H)(L) ← (H)(L) + 1

Program example 4.1: addition instructions

The program example shown in Figure 4.1 illustrates the effect of the previously described addition instructions on two two's complement signed

(a) Flowchart

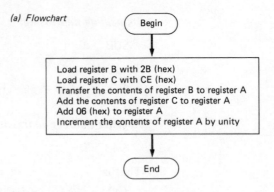

(b) Program code

Symbolic instructions	Action
LD B,2B	(B) ← 2B (hex), i.e. $+43_{10}$
LD C,CE	(C) ← CE (hex), i.e. -50_{10}
LD A,B	(A) ← (B), i.e. (A) ← 43_{10}
ADD A,C	(A) ← F9 (hex), i.e. -7_{10}
ADD A,06	(A) ← FF (hex), i.e. -1_{10}
INC A	(A) ← 00 (hex), i.e. 0_{10}

Figure 4.1   Program example 4.1

values which are loaded into registers B and C. As has been mentioned, the microprocessor simply performs the bit-by-bit addition operations as specified in the instructions and it is the user who interprets the resulting binary patterns as representing meaningful two's complement signed values.

Program example 4.2:   subtraction instructions

The program example shown in Figure 4.2 illustrates the effect of the previously described subtraction instructions on two two's complement signed values which are first loaded into registers B and C. Again ensure the signs of the numbers are interpreted in the correct way after each subtraction operation.

Program example 4.3:   mixed arithmetic

The program example shown in Figure 4.3 illustrates the effect of addition and subtraction instructions on two two's complement signed values. Perform the bit-by-bit addition and subtraction operations to verify the results shown in the actions.

Q4.3(b)

4.4   The flags register

The microprocessor is a versatile device since it is applied to perform a specific application function after it has been manufactured. The user selects the appropriate sequence of instructions from the instruc-

(a) Flowchart

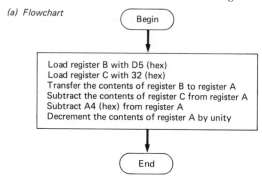

(b) Program code

Symbolic instructions	Action
LD B,D5	$(B) \leftarrow D5$ (hex), i.e. -43_{10}
LD C,32	$(C) \leftarrow 32$ (hex), i.e. $+50_{10}$
LD A,B	$(A) \leftarrow (B)$, i.e. $(A) \leftarrow -43_{10}$
SUB C	$(A) \leftarrow A3$ (hex), i.e. -93_{10}
SUB A4	$(A) \leftarrow 01$ (hex), i.e. 1_{10}
DEC A	$(A) \leftarrow 00$ (hex), i.e. 0_{10}

Figure 4.2 Program example 4.2

(a) Flowchart

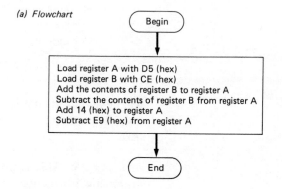

(b) Program code

Symbolic instructions	Action
LD A,D5	$(A) \leftarrow D5$ (hex), i.e. -43_{10}
LD B,CE	$(B) \leftarrow CE$ (hex), i.e. -50_{10}
ADD A,B	$(A) \leftarrow A3$ (hex), i.e. -93_{10}
SUB B	$(A) \leftarrow D5$ (hex), i.e. -43_{10}
ADD A,14	$(A) \leftarrow E9$ (hex), i.e. -23_{10}
SUB E9	$(A) \leftarrow 00$ (hex), i.e. 0_{10}

Figure 4.3 Program example 4.3

tion set available with the selected device (the program) and this is then executed to perform the required application function.

Since different users may require to interpret the binary data stored within the microcomputer in different ways, the microprocessor manufacturers incorporate a range of facilities into the device so that the maximum flexibility in its use can be achieved. One such feature is the provision of *the flags (or status) register* in the arithmetic logic unit.

The flags register is simply a collection of individual status or condition bits. Each bit is automatically either set to logical 1 or reset to logical 0, according to certain rules, whenever an arithmetic instruction is executed. The programmer can then inspect or utilise instructions which interpret the condition of selected flags in order to manipulate the data stored within the microcomputer in the required way.

As an example, some of the individual flag bits which comprise the flags register of the Z80 are shown in Figure 4.4. The use of these flags

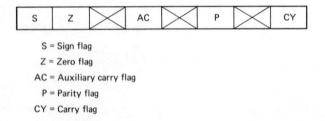

S = Sign flag
Z = Zero flag
AC = Auxiliary carry flag
P = Parity flag
CY = Carry flag

Figure 4.4   Some Zilog Z80 flags

will be discussed later in this and the next chapter, but the effect on these flags of the previously discussed arithmetic instructions is given here.

Sign flag, S   This is set to logical 1 whenever the result of an arithmetic instruction is negative, i.e. in the two's complement notation whenever the most significant bit of register A is a binary 1. Otherwise the flag is left reset after the instruction.

Zero flag, Z   This is set to logical 1 whenever the result of an arithmetic instruction leaves zero in register A, otherwise the flag is reset.

Auxiliary carry flag, AC   This is used when BCD number representation is being used. This will be discussed later in this chapter. The flag is set when the result of an arithmetic operation produces a carry out from the least significant half of register A.

Parity flag, P This is used with the logical operations to be discussed later in this chapter. The flag is set if the result of a logical instruction produces an even number of ones in register A, otherwise the flag is reset.

Carry flag, CY This is in fact the carry out from the most significant bit of register A. It is therefore set after an addition or subtraction instruction whenever a carry or a borrow is generated from the most significant bit of register A. One of its main uses will now be described.

4.5 The carry flag

The majority of addition and subtraction instructions available with an 8-bit microprocessor perform arithmetic operations on 8-bit values. Thus, since these are used to represent two's complement signed binary numbers, the maximum range of number that can be accommodated in an 8-bit system is from +127 to –128, inclusive.

For many microprocessor applications this is quite adequate since the amount of actual processing performed is often quite small. In some applications, however, this is not sufficient and the user has then to arrange, firstly, to allow more than a single byte of the memory to hold or store each value and, secondly, when processing these values, to do so in a number of steps.

Assume in an application it is required to represent signed numbers in the range +32,767 to –32,768, i.e. each number requires 16 bits or 2 bytes, and that the microprocessor being used does not contain any 16-bit arithmetic instructions.

Firstly, each number stored within the microcomputer memory must be allocated two bytes:

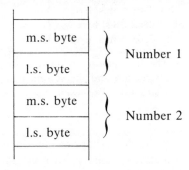

Secondly, since the microprocessor can only add or subtract two 8-bit values with a single instruction, each addition or subtraction operation must be performed in two steps: first the least significant pair of bytes must be added/subtracted and then the most significant pair. In addition, however, it can be seen that when adding/subtracting the second (most significant) pair of bytes it is necessary to include any carry/borrow that may have been generated when the first (least significant) pair of bytes were being processed.

It is for this reason that microprocessors normally provide two alternative addition and subtraction instructions: the first as have been described and the second which involve the carry (borrow) bit. For example, with the Z80:

    ADC   A,B

This results in the contents of register B *and* the current carry bit (flag) being added to the current contents of register A. Thus in the above example, if the addition of the first pair of bytes produces a carry out then this can be added with the next pair of bytes by using the above add with carry (ADC) instruction. That is,

    (A) ← (A) + (B) + (CY)

Similarly, the instruction

    SBC   A,B

results in the contents of register B *and* the contents of the carry (borrow) flag being subtracted from the contents of register A, i.e.

    (A) ← (A) - (B) - (CY)

Program example 4.4:   addition with carry

The program example shown in Figure 4.5 illustrates the use of the add with carry instruction. The program adds together two 16-bit (2-byte) numbers which are stored in four consecutive memory locations as follows:

    2800 = l.s. byte }
    2801 = m.s. byte } Number 1
    2802 = l.s. byte }
    2803 = m.s. byte } Number 2

The first pair of bytes are first added together using the normal ADD instruction and the second pair of bytes added together using the ADC instruction. The result replaces the first number in memory.

It should perhaps be mentioned here that the Z80 does in fact contain a limited number of 16-bit (2-byte) arithmetic instructions and hence the above example could be implemented with fewer instructions if these were used. The aim of the example, however, is to illustrate how multiple-byte numbers can be processed with an 8-bit microprocessor. The approach adopted is readily applicable to numbers of any length.

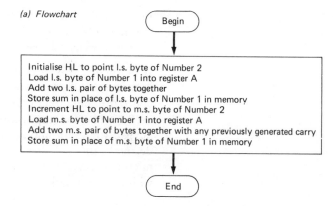

(a) Flowchart

Begin

Initialise HL to point l.s. byte of Number 2
Load l.s. byte of Number 1 into register A
Add two l.s. pair of bytes together
Store sum in place of l.s. byte of Number 1 in memory
Increment HL to point to m.s. byte of Number 2
Load m.s. byte of Number 1 into register A
Add two m.s. pair of bytes together with any previously generated carry
Store sum in place of m.s. byte of Number 1 in memory

End

(b) Program code

Symbolic instructions	Action
LD HL,2802	(H)(L) points to l.s. byte of Number 2
LD A,(2800)	(A) ← l.s. byte of Number 1
ADD A,(HL)	l.s. pair of bytes added together
LD (2800),A	l.s. byte of sum replaces l.s. byte of Number 1
INC HL	(H)(L) points to m.s. byte of Number 2
LD A,(2801)	(A) ← m.s. byte of Number 2
ADC A,(HL)	m.s. pair of bytes and carry added together
LD (2801),A	m.s. byte of sum replaces m.s. byte of Number 1

Figure 4.5 Program example 4.4

4.6 Binary-coded decimal arithmetic

The previously discussed arithmetic instructions are intended for performing arithmetic operations on data which are stored in their pure binary form. As was mentioned in Chapter 1, it is necessary to convert all information input to the microcomputer into this form prior to processing and also, for output, to convert the internally stored binary results into the required output form.

As has been indicated, the data processed by a microcomputer can be from a variety of different sources. For example, if the input and subsequent output are continuously varying analogue signals, then an analogue-to-digital converter (ADC) is used to convert the input analogue signal into its binary form. Also, a digital-to-analogue converter (DAC) is used to convert the binary values output by the microcomputer into their analogue form.

Similarly, if the input data is from a keyboard the binary-coded data produced at the mirocomputer input port (from an associated key-board encoding circuit) must be converted into its pure binary form prior to processing by the microprocessor. Hence, although this is not a difficult task and can in fact be performed by a small piece of program code (list of instructions), if the eventual amount of processing to be performed on the subsequent data is small, a considerable proportion of the processing effort of the microprocessor can be spent doing the conversion process.

It is for this reason that most microprocessors also provide some limited facilities for processing data which are stored in a binary-coded decimal rather than a pure binary form. These are particularly useful when the application uses, for example, a simple numeric keypad for input and for output, a display comprised of a row of decimal digits. The data entered and stored within the microcomputer can then be processed and subsequently output without any intermediate conversion.

Table 4.5

Decimal digit	BCD code
0	0000
1	0001
2	0010
3	0011
4	0100
5	0101
6	0110
7	0111
8	1000
9	1001

BCD representation

The output from a simple (decimal) numeric keypad is normally a 4-bit binary-coded value. Since this value represents a single decimal digit (0–9), it is referred to as a *binary-coded decimal or BCD* value. BCD representation is in fact a subset of the hexadecimal system introduced in Chapter 1 and hence a list of the 4-bit equivalent BCD codes of the ten decimal digits are as shown in Table 4.5.

Since each decimal digit using BCD representation requires 4 bits, an 8-bit binary value stored within a microcomputer can represent two decimal digits. Some examples of stored 8-bit patterns and their equivalent decimal numbers are therefore as follows:

$$0101\ 0111 = 57\ \text{(decimal)}$$
$$1000\ 1001 = 89\ \text{(decimal)}$$
$$0001\ 0000 = 10\ \text{(decimal)}$$
$$0010\ 0100 = 24\ \text{(decimal)}$$

BCD addition

It can be deduced from the above discussion that an 8-bit binary value stored within the microcomputer memory may represent either a two's complement signed binary value in the range −128 to +127 or two BCD digits in the range 00 to 99. All the arithmetic instructions previously discussed, however, assume the data are stored in their two's complement binary form. When manipulating stored values which represent BCD digits, therefore, it is necessary to perform some adjustments to the normal binary results in order to produce the correct results in the BCD form. This is perhaps best illustrated by means of some examples.

Consider the addition of two 8-bit binary values, A and B, each of which represents two BCD digits:

$$A = 0011\ 0100 = 34\ \text{BCD}$$
$$B = 0010\ 0001 = 21\ \text{BCD}$$
$$A + B = 0101\ 0101$$

Correct BCD
    sum   = 0101 0101 = 55 BCD

It can be seen from this example that the binary sum produced by the normal bit-by-bit addition operations does in fact yield the correct result in its BCD form. Consider, however, the following addition:

$$A = 0111\ 0110 = 76\ \text{BCD}$$
$$B = 0001\ 0101 = 15\ \text{BCD}$$
$$A + B = 1000\ 1011$$

Correct BCD
    sum   = 1001 0001 = 91 BCD

Clearly, the sum produced by the normal bit-by-bit addition operation is now different from the correct BCD sum required. This is so because the sum produced by the normal binary addition yields a result which is in excess of nine in the least significant four bits. Similarly, consider the addition:

$$A = 0100\ 1000 = 48\ \text{BCD}$$
$$B = 0100\ 1001 = 49\ \text{BCD}$$
$$A + B = 1001\ 0001$$

Correct BCD
    sum   = 1001 0111 = 97 BCD

Again, the normal binary sum is incorrect because although the bit-by-bit addition operation naturally produces a carry out from the least significant four bits, the resulting four bits are still incorrect.

In practice it is in fact a reasonably straightforward task to produce the corrected BCD sum from the result produced by the normal binary addition operation. The procedure can readily be deduced from the above examples and is summarised as follows:

> If the normal binary addition of each 4-bit group produces a result which is less than 9 *and* a carry is not produced when adding the two least significant BCD digits, then the result is correct. Else, if the result of the normal binary addition is greater than 9 *or* a carry is generated when adding the two least significant BCD digits, then a correction of +6 must be added to the normal binary sum.

To illustrate the effect of this procedure consider the previous examples. Since in the first example the result produced in each 4-bit group is less than 9, the normal binary sum yields the correct result and no correction is required. In the second example, however, the sum produced when adding the two least significant BCD digits is greater than 9 and in the third example a carry is generated when the two least significant digits are being added. The corrected sums for the last two examples can therefore be derived as follows:

1   Uncorrected sum (A + B)   =   1000 1011
    Add correction (+6)   +0000 0110

    Corrected BCD sum   =   1001 0001 = 91 BCD

2   Uncorrected sum (A + B)   =   1001 0001
    Add correction (+6)   +0000 0110

    Corrected BCD sum   =   1001 0111 = 97 BCD

As was indicated in Section 4.4, one of the flags associated with the flags register is known as the auxiliary carry flag, AC. This flag is in fact provided for use specifically when BCD number representation is being used since it is the carry out from the least significant 4-bit group. In addition, an instruction is normally provided in a micro-processor instruction set to perform the above correction auto-matically. When executed the instruction performs the above addition operation depending on either the state of the AC flag or the result produced after the normal addition process. In the Z80 it is called the *d*ecimal *a*djust *a*ccumulator or DAA since the result of the normal binary addition will automatically be left in the accumulator.

The DAA instruction is, in fact, designed to correct both halves of the normal 8-bit binary sum, i.e. both the least significant and the most significant 4-bit groups. As with the previous discussion concerned with the carry flag and the addition of multiple-byte numbers, this provides the facility for processing BCD numbers comprised of more than two digits. The effect of the DAA instruction is illustrated in the following program example.

Program example 4.5: BCD addition

The program example of Figure 4.6 adds together three 8-bit numbers. Each number is first loaded into a microprocessor register and represents two BCD digits. Notice that after each normal binary addition, the resulting 8-bit value is corrected by means of the DAA instructions. The final result is in register A.

BCD subtraction

The procedure to produce the correct BCD difference from the result produced by the normal binary subtraction operation is unfortunately different from that which has just been described for addition. In fact the corrections required are far more complicated as can be seen from the table of corrections shown in Table 4.6. The table shows the state of the contents of register A and the two carry flags *after* the normal binary subtraction instruction has been executed. Consider the following example

$$A = 0111\ 0110 = 76\ \text{BCD}$$
$$B = 0010\ 1000 = 28\ \text{BCD}$$
$$A - B = \overline{0100\ 1110} \quad CY = 0,\ AC = 1$$

Referring to the figure of corrections CY is 0, the most significant 4-bit group is 4, AC is 1 and the least significant 4-bit group is E (hex). The correction to be *added* is therefore FA (hex):

A – B	=	0100 1110
Add correction		+1111 1010 = FA (hex)
Corrected BCD difference	=	0100 1000 = 48 BCD

The majority of decimal-adjust instructions provided with microprocessors only perform the correct adjustment after the addition operation and hence with these devices it is the responsibility of the

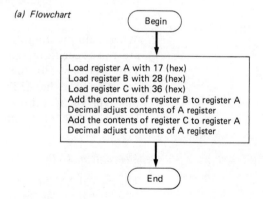

(a) Flowchart

Begin

Load register A with 17 (hex)
Load register B with 28 (hex)
Load register C with 36 (hex)
Add the contents of register B to register A
Decimal adjust contents of A register
Add the contents of register C to register A
Decimal adjust contents of A register

End

(b) Program code

Symbolic instructions	Action
LD A,17	(A) ← 17 (hex), i.e. 17 BCD
LD B,28	(B) ← 28 (hex), i.e. 28 BCD
LD C,36	(C) ← 36 (hex), i.e. 36 BCD
ADD A,B	(A) ← 3F (hex)
DAA	(A) ← 45 (hex), i.e. 45 BCD
ADD A,C	(A) ← 7B (hex)
DAA	(A) ← 81 (hex), i.e. 81 BCD

Figure 4.6 Program example 4.5

Table 4.6   *Corrections to produce BCD difference*

Carry (CY)	Most significant 4-bit group	Auxiliary carry (AC)	Least significant 4-bit group	Correction to be added
0	0 – 9	0	0 – 9	00 (hex)
0	0 – 8	1	6 – F	FA
1	7 – F	0	0 – 9	A0
1	6 – F	1	6 – F	9A

user (programmer) firstly to deduce the type of correction required according to the rules tabulated in Table 4.6 and then to perform the appropriate correction. Only in a very few microprocessors (the Z80 for example) does the decimal-adjust instruction perform the appropriate correction after both addition and subtraction operations. The Z80 effectively remembers whether the previous arithmetic instruction was an addition or subtraction and performs the appropriate correction when the DAA instruction is executed.

Q4.4–4.6

4.7   Logic instructions

Unlike electronic calculators, many microprocessor applications are not concerned with numeric values. Very often an 8-bit value read from an I/O interface port does not represent a number as such but rather the state of, say, eight digital devices.

When manipulating this data, therefore, some common operations are 'test a specific bit (and hence device) to determine if it is on (logical 1) or off (logical 0)'; 'determine if any bits (and hence device states) in the value have changed'; and so on. The instructions used to perform this type of operation are the logical instructions AND, OR and Exclusive-OR. Each will be considered separately.

The logical AND

The logical AND instruction is particularly useful for performing the so-called *masking operation*. This is an operation which effectively *masks out* unwanted bits in an 8-bit group and hence provides a mechanism, for example, for testing the state of, say, a specific bit in the group.

Table 4.7   *The logical AND operation*

Bit 1	Bit 2	Bit 1 AND Bit 2
0	0	0
0	1	0
1	0	0
1	1	1

A truth table showing the logical AND operation on two binary digits is shown in Table 4.7. It can readily be deduced from the figure that the output is a logical 1 only when Bit 1 AND Bit 2 are both 1. Otherwise the output is logical 0.

The logical AND instruction in a microprocessor performs the equivalent of eight AND operations in parallel since it performs the bit-by-bit logical AND operation between the contents of register A and either the contents of another microprocessor register or memory location or an immediate data value.

For example, assuming the contents of registers A and B are B3 (hex) and D8 (hex), respectively, the result of performing the instruction:

AND   B

would be as follows:

$$(A) = 1011\ 0011 = B3\ (hex)$$
$$(B) = 1101\ 1000 = D8\ (hex)$$

$$(A)\ AND\ (B) = \overline{1001\ 0000} = new\ contents\ of\ A$$

First the least significant pair of bits are ANDed together, then the next pair and so on. Notice that only when both values contain a logical 1 in the same column is there a 1 in the output.

Similarly, assuming the contents of register A are 6B (hex), the result of performing the instruction

AND   08

would be:

$$(A) = 0110\ 1011 = 6B\ (hex)$$
$$Immediate\ data = 0000\ 1000 = 08\ (hex)$$

$$(A)\ AND\ 08 = \overline{0000\ 1000} = new\ contents\ of\ A$$

Notice in this example that wherever there is a 0 in the immediate data value there is automatically a 0 in that bit position in the result. Where there is a 1, however, then the output will be determined by the state of the corresponding bit in A: if the bit is a 1 (as in the example) the output will be 1, if the bit is a 0 the output will be 0.

It can be deduced from this, therefore, that the state of a specific bit in an 8-bit value can be tested, first by performing the logical AND operation between the value and an immediate data value which has a logical 1 in the bit position to be tested. Then, if the resulting contents of register A are zero, the bit was a 0 else, if the contents are non-zero (as in the example), the bit was a 1.

The logical instructions also affect the individual flag bits in the flags register and consequently, after performing the logical AND operation, it is usual to follow this with an instruction whose action is conditional on the state of, say, the zero flag. These are known as the conditional branch or jump instructions and will be discussed in the next chapter.

The logical OR

When performing certain processing operations it is sometimes necessary to form a value which is made up of, say, the four most significant bits of one value and the four least significant bits of a second value. To perform this operation, first the four least significant bits are masked out from the first value and the four most significant bits masked from the second (both using the logical AND instruction) and then the two results are *merged* together using the logical OR instruction. This will be illustrated after the latter has been explained.

A truth table showing the logical OR operation is shown in Table 4.8. It can be seen that the output is a logical 1 when either Bit 1 *OR* Bit 2 is a 1 *OR* when both are a 1. Otherwise the output is logical 0.

As with the logical AND instruction, when executing the logical OR the microprocessor performs the equivalent of eight OR operations on each such successive pair of bits. The first value is held in register A and the second value is either the current contents of a microprocessor register or memory location or an immediate data value specified in the instruction.

For example, if the contents of the registers A and B are 63 (hex) and 0F (hex), respectively, the instruction

    OR   B

results in the following:

$$(A) = 0110\ 0011 = 63\ \text{(hex)}$$
$$(B) = 0000\ 1111 = 0F\ \text{(hex)}$$
$$(A)\ \text{OR}\ (B) = \overline{0110\ 1111} = \text{new contents of A}$$

Notice in this example that where the bits in the second value (B) are 1 the output is always 1 but where the bits in the second value are 0, the outputs are determined by the state of the corresponding bits in the first value (A). If they are also 0 the outputs are 0; if they are 1 the outputs are 1.

Similarly, the instruction

    OR   F0

assuming the contents of A are the same, would be:

$$(A) = 0110\ 0011 = 63\ \text{(hex)}$$
$$\text{Immediate data} = \overline{1111\ 0000} = F0\ \text{(hex)}$$
$$(A)\ \text{OR}\ F0 = \overline{1111\ 0011} = \text{new contents of A}$$

To illustrate the merging operation mentioned earlier, consider the following program example which uses a combination of both the logical AND and OR instructions.

Table 4.8   *The logical OR operation*

Bit 1	Bit 2	Bit 1 OR Bit 2
0	0	0
0	1	1
1	0	1
1	1	1

Program example 4.6: Logical AND and OR instructions

The program example of Figure 4.7 illustrates a typical use of the logical AND and OR instructions. The program generates an 8-bit value comprised of the most significant four bits of the contents of register B and the least significant four bits of the contents of register C. The unwanted bits of each

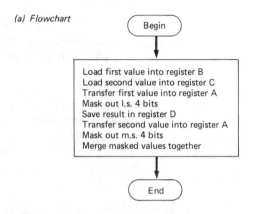

(a) Flowchart

Begin

Load first value into register B
Load second value into register C
Transfer first value into register A
Mask out l.s. 4 bits
Save result in register D
Transfer second value into register A
Mask out m.s. 4 bits
Merge masked values together

End

(b) Program code

Symbolic instructions	Action
LD B,6A	(B) ← 6A (hex)
LD C,37	(C) ← 37 (hex)
LD A,B	Load first value in A
AND F0	Mask out l.s. four bits
LD D,A	Save result in register D
LD A,C	Load second value in A
AND 0F	Mask out m.s. four bits
OR D	Merge masked values together, i.e. (A) ← 67 (hex)

Figure 4.7 Program example 4.6

value are first masked out using the logical AND instruction and the required result produced by merging the two masked values together using the logical OR instruction. The result is left in register A.

The Exclusive OR

Assuming an 8-bit value read from an I/O interface port represents the state of eight digital devices, a frequent processing task is to continuously monitor the state of these devices and only initiate a further processing action when one of these devices changes state, 1 to 0 (on to off) or 0 to 1 (off to on). The instruction used to perform this task is the Exclusive OR or XOR instruction.

The XOR operation is shown in the truth table in Table 4.9. It can be seen that the output is the same as the logical OR operation except that with the XOR the output is 1 only when the two input bits are different, 0,1 or 1,0, and the output is 0 when the two input bits are the same, 0,0 or 1,1.

Table 4.9 *The logical XOR operation*

Bit 1	Bit 2	Bit 1 XOR Bit 2
0	0	0
0	1	1
1	0	1
1	1	0

Again, the XOR operation involves the contents of register A and either the contents of another register or memory location or an immediate data value. Hence if the current contents of register A are B3 (hex), the effect of the instruction

```
XOR   AC
```

would be:

$$(A) = 1011\ 0011 = B3\ (hex)$$
$$\text{Immediate data} = 1010\ 1100 = AC\ (hex)$$
$$(A)\ XOR\ B3 = \overline{0001\ 1111} = \text{new contents of A}$$

To illustrate an application of the XOR instruction, assume the current state of eight controlled devices (on or off) is held in an 8-bit value stored within register B and the new state of the devices is read into register A. A single XOR instruction between the two values would then indicate if the two values are different and hence if the state of any of the devices has changed. This can be achieved using the instruction:

```
XOR   B
```

and an example of its effect is as follows:

$$(A) = 1010\ 0110 = \text{new state of devices}$$
$$(B) = 1110\ 0110 = \text{old state of devices}$$
$$(A)\ XOR\ (B) = \overline{0100\ 0000} = \text{zero if no devices changed, non-}$$

zero if one or more devices have changed

As with the other logical instructions, the XOR instruction affects the various flag bits and consequently this would normally be followed by an instruction whose action is conditional on the state of the zero flag: if it is set, no changes have taken place and hence continue monitoring; if it is not set, a change has occurred and hence perform some actions in response to the change. This will be explained in the next chapter.

Q4.7

4.8   Shift instructions

In addition to the logical instructions just presented, a micro-processor also contains instructions to perform bit manipulation or shift operations on binary values. The shift instructions have a number of uses. For example, if after performing the XOR instruction between two values (as was described in the previous section) a change of state was detected, it is then necessary to determine the specific bit (and hence device) which has changed so that the appropriate action can be carried out. This can conveniently be done using the shift instructions.

The shift operation involves shifting the contents of register A either left or right one place. In addition, the shift instructions on a microprocessor normally involve the carry flag (CY) so that subsequent tests can be made on its status.

For example, consider the instruction

RLCA

This results in the contents of register A being shifted left one place. The most significant bit of register A is effectively looped around to become the new contents of the least significant bit of the register and also the new contents of the carry flag:

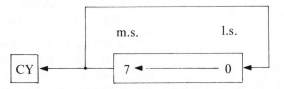

As an example, assume the current contents of register A are A6 (hex). Then the result of the RLCA instruction would be:

(A) = ┌─10100110 = A6 (hex)

1 01001101 = 4D (hex)

That is, the new contents of A would be 4D (hex) and the carry flag would be set (1). Another RLCA instruction would then have the following effect:

(A) = ┌─01001101 = 4D (hex)

0 10011010 = 9A (hex)

That is, the new contents of A would be 9A (hex) and the carry flag would now be reset (0). Clearly, if the current contents of A held the result of the XOR operation mentioned earlier, then repeated execution of RLCA instructions until the carry flag becomes set would be a mechanism for determining which bit was set and hence had changed.

In addition, the shift instructions are widely used when performing binary multiplication and division operations. As was mentioned earlier, most 8-bit microprocessors do not contain a multiplication or division instruction and hence, if these operations are required, the programmer must use a list of instructions similar to those just described in order to perform these functions. Binary multiplication, for example, can be performed using repeated left shifts since a single left shift operation is equivalent to a times two operation. Similarly, binary division can be performed by repeated right shifts since each right shift is a divide by two operation.

Q4.8

4.9   The compare instruction

A very common and useful programming operation (as will be seen in the next chapter) is to repeatedly execute an instruction or list of instructions a specific number of times. This is accomplished by performing a test after each pass through the list to determine if the appropriate number of passes have been performed. If it has, execution continues in sequence. If it hasn't, a branch back to the start of the list is performed.

The instructions which have so far been described all modify the contents of register A. Thus if, let us say, we wish to perform a selected operation ten times, we may typically do this using one of the microprocessor registers as a counter. The contents of the counter are incremented by unity each time the operation is performed and a test carried out to determine if the contents have reached ten. If they have not, the operation is performed again but if they have, the operation is terminated.

Hence when carrying out the test operation it is essential that the contents of the register (counter) are not destroyed. Clearly a simple subtract instruction would destroy the current contents (count) and hence a more convenient method is to use the compare instruction.

The compare instruction compares two values without modifying either value. It does, however, affect the various flags – the zero flag, for example, is set only if the two values are the same. The two values compared are the contents of register A and either the contents of another microprocessor register or memory location or an immediate data value. Thus the instruction

```
CP    0A
```

would compare the current contents of register A with the immediate data value 0A (hex) or 10 decimal. If they are equal, the zero flag is set; if they are not, the zero flag is reset. Hence if register A held the count mentioned earlier, then a simple test operation on the state of the zero flag could be used to determine when the looping operation should be terminated. This technique is used extensively in the next chapter.

Questions

4.1 Express the following decimal numbers in their 8-bit two's complement signed binary form:

+63   +112   +37
−111   −27   −78

4.2 Derive the decimal equivalent of the following binary numbers, assuming two's complement signed number representation is being used:

01011011   00010111   01111111
11110111   10000101   11010101

4.3 Convert the following decimal numbers into their 8-bit two's complement signed binary form and perform the indicated addition or subtraction operation. Convert your answers into their decimal forms to verify these are correct.

(a)   65 + 24   −28 + 81   −63 + (−56)
(b)   94 − 36   −53 − 23   −68 − (−84)

4.4 Assuming BCD number representation, use the corrections derived in Section 4.6 to find the BCD results of the following addition and subtraction operations. Convert the numbers into their decimal form to verify the results.

01100111 + 00100100
10001001 + 00001000
01110100 − 01001000
10011000 − 00111001

4.5 Assuming BCD number representation, derive the decimal equivalent of the binary number left in register A after the following instructions have been executed:

```
LD   A,27
ADD  A,56
DAA
SUB  38
DAA
```

4.6 Write a program to first load the three numbers shown in the following sum into registers B, C and D, respectively, and then perform the indicated operations assuming BCD number representation:

26 + 69 − 56

4.7 Derive the result of performing the logical AND, OR and XOR operations on the following pairs of hexadecimal numbers:

A7, B8
E2, DC
3C, 73
5A, 6B

4.8 Determine the hexadecimal number left in register A and the state of the carry flag after the following program has run:

```
LD      A,6F
LD      B,A4
AND     B
OR      44
RLCA
RLCA
```

Chapter 5 Transfer of control instructions

Objectives of this chapter *When you have completed studying this chapter you should be able to:*

1   *Understand the structure of a typical unconditional and conditional jump instruction.*
2   *Describe the actions performed by a microprocessor when implementing these instructions.*
3   *Write and understand programs that use jump instructions and appreciate the advantage of using symbolic address labels.*
4   *Use the symbols employed in the flowchart method of program design and be able to design simple programs using them.*
5   *Understand the terms* looping *and* binary decisions *and be able to design and implement simple programs which use them.*
6   *Recognise the importance of the use of subroutines in program design.*
7   *Describe the actions performed by a microprocessor when implementing a subroutine CALL and RETurn instruction.*
8   *Describe the structure of a stack and how its contents are affected by the various microprocessor instructions which use it.*
9   *Describe the term* parameter *and be able to write short programs which include passing parameters to a subroutine by means of a memory block.*
10 *Describe the term* nested subroutines *and understand its implications on the design of a subroutine.*

5.1   Introduction

All the instructions so far discussed perform a specific data movement or data manipulation operation. Thus if this was the only type of instruction available, when the microprocessor executed a program it would simply start at the first instruction and execute the remaining instructions sequentially until it reached the end of the list, i.e. when executing a program, the microprocessor would execute each instruction in the program just once only.

The main advantage of the stored-program concept, however, stems from the fact that once a program has been stored in memory, the same list of instructions can be executed many times over. This can be done by the operator re-running the program from the command console, of course, but also, and more important, it is possible for the microprocessor itself to automatically re-execute the same list or a group of instructions many times over during a single running of the program.

Hence to provide this facility, a microprocessor contains, in addition to instructions which perform a specific data movement or arithmetic operation, instructions that can control or modify the actual sequence of program execution. These are known as *transfer of control* instructions and, as will be seen, they provide the programmer with considerable flexibility when writing a program. Not only are instructions provided to branch back to a specific part (instruction) of the program, but also instructions to make the branching operation conditional on certain conditions being satisfied during actual program execution. Both types will be considered.

5.2   The unconditional jump instruction

All microprocessors provide an unconditional jump instruction. This is the facility that enables the programmer to break the normal sequential mode of execution of a program and branch unconditionally to a different part of the program to obtain the next instruction before continuing its sequential action once again.

An unconditional jump instruction is comprised, therefore, of an operation specifying the instruction type and an address part which, instead of specifying, say, the address in memory which contains a value to be manipulated, specifies the address in memory which contains the next instruction to be executed.

Table 5.1   *The action of the unconditional jump instruction*

Memory addresses	Symbolic instructions	Action
2000	–	
.	.	
.	.	
.	.	
2010	– ◄───	Destination instruction
.	.	
.	.	
.	.	
2020	JP,2010 ─┘	Branch to address 2010 (hex)
.	.	
.	.	
.	.	
2030	–	

For example, with the Z80 the unconditional jump instruction is written as

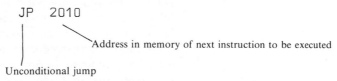

JP   2010

Unconditional jump

Address in memory of next instruction to be executed

Hence, assuming the list of instructions which comprise a program is stored in memory starting at address 2000 (hex) and the above jump instruction is stored at 2020 (hex), the effect of executing this instruction is illustrated in Table 5.1.

The unconditional jump instruction is normally a three-byte instruction and for the Z80 the above instruction is equivalent to:

Memory
address   *Contents (hex)*

Address	Contents	
2020	C3	= JP or unconditional jump
2021	10	= l.s. byte of destination address
2022	20	= m.s. byte of destination address

Hence, when executing this instruction, during the fetch phase of the instruction cycle the microprocessor reads the first byte of the instruction [C3 (hex)] from the address in memory currently held in the program counter [2020 (hex)] and loads this into the instruction register. It then determines from this that the instruction is an unconditional jump and hence two further bytes must be read from memory. The next two sequential bytes are therefore fetched from memory and the contents of the program counter are incremented by one after each byte has been read.

Hence, after three bytes of the instruction have been read from memory, the program counter will automatically contain the address in memory of the first byte of the next sequential instruction [2023 (hex)]. During the execution phase of the instruction cycle, however, the microprocessor loads the two address bytes read from memory during the fetch phase [20 and 10 (hex)] into the program counter, thereby overwriting the current contents.

The microprocessor then proceeds to the next fetch phase and hence reads the first byte of the next instruction from the current address held in the program counter. Instead of this pointing to the first byte of the next sequential instruction, however, it now points to address 2010 (hex), i.e. to the first byte of the out-of-sequence instruction. Program execution will then continue sequentially until another transfer of control instruction is executed.

Instruction labels

The unconditional jump instruction can be used to branch or jump either back to an earlier instruction as was indicated above or forward to a later instruction in the program. The usefulness of the latter will be illustrated later in this chapter but it can be deduced from this, that when the programmer is writing a program containing a jump instruction, he does not always know the absolute address in memory of where the destination instruction is stored.

When writing a program that involves jump instructions, therefore, it is usual for the programmer to use *labels* or *symbolic address names* to indicate the intended addresses of the destination instructions and it is only during the translation process that these are converted into absolute addresses. This is illustrated in Table 5.2, which contains two unconditional jump instructions: one to cause a jump out of sequence back to the instruction which is given the symbolic address name (or label) LAB1 and the other to jump forward out of sequence to the instruction with the symbolic address label LAB2.

Normally, the symbolic names used for labels are a mix of alphabetic

Table 5.2 *The use of labels as symbolic address names*

Memory addresses	Symbolic instructions	Action
Assigned during translation process	LAB1: Destination instruction	
	JP LAB1	Branch to instruction with symbolic address label LAB1
	JP LAB2	Branch to instruction with symbolic address label LAB2
	LAB2: Destination instruction	

and numeric characters and are usually made meaningful names by the programmer. For example,

```
SUM:
ALARM:
REPEAT:
END:
```

are all examples of labels whose meaning can be readily deduced. This, as will be seen, can substantially improve the readability and hence understandability of a program containing transfer of control instructions.

Q5.1

5.3 Conditional jump instructions

Although the unconditional jump instruction provides a useful function, the real flexibility of the stored-program concept comes from the provision of the conditional type of jump instruction. Thus, instead of the sequential order of program execution being broken unconditionally, it is broken only if a specified condition is met, otherwise normal sequential operation is maintained.

To illustrate the usefulness of the conditional jump instructions consider writing a program segment to, say, form the sum of one hundred values which are stored in one hundred consecutive memory locations. These may typically be values which have been read from an I/O interface port by an earlier group of instructions in the program.

Without the use of a jump instruction this would clearly necessitate a list of one hundred repetitive operations which simply read each value in sequence and add the value to the current running total held in, say, the accumulator. Thus, assuming two instructions per addition operation – one to prepare the address of the next value in memory and the other to perform the addition – the program segment would require two hundred instructions as follows:

```
      .
      .
      .
LD    A,00       Initialise total to zero
LD    HL,2800    Initialise HL to point to first
                   value
ADD   A,(HL)     Add first value to running total
INC   HL         Increment HL to point to second
                   value
ADD   A,(HL)     Add second value to running
                   total
```

(Continued)

```
INC   HL          Increment HL to point to third
                    value
ADD   A,(HL)      Add third value to running total
       .
       .
       .
INC   HL          Increment HL to point to 100th
                    value
ADD   A,(HL)      Add 100th value to running total
       .
       .
       .
```

It can readily be deduced from the above that after the accumulator and the address pointer (H,L) have been initialised, the program is comprised simply of a list of repetitive operations. Hence, if instead of the second addition instruction a jump instruction is used to cause the microprocessor to branch back to the first addition instruction, then each load and addition operation would be carried out repetitively and hence the total program segment need only then be comprised of five instructions.

The problem with this solution is that if an unconditional jump instruction is used, the microprocessor will not know when all the values have been summed together and will simply continue branching back or *looping* continuously. This can readily be over- come, however, by using a conditional jump instruction. As will be seen, by maintaining a record of the number of times the branch operation has been carried out – the loop count – it is possible to determine when this has reached a preset limit. The branching action then ceases and normal sequential program execution resumed.

Jump conditions

The conditions specified with the conditional jump instructions are all determined by the state of the various flag bits present in the microprocessor flags register. For example, there are normally two conditional jump instructions based on the state of the zero or Z flag. With one, a jump (branch) only occurs if the zero flag is set whilst with the other, a jump only occurs if the zero flag is not set. Then, depending on which one is used, if the appropriate condition is not satisfied a jump will not occur and normal sequential execution will follow.

As an example, some of the jump conditions that can be specified using the Z80 are shown in Table 5.3. The instructions listed are based

Table 5.3 *Conditional jump mnemonics*

Instruction menmonic	Flag status	Action
JP Z	Z = 1	Jump if zero flag set
JP NZ	Z = 0	Jump if zero flag not set
JP C	C = 1	Jump if carry (flag) set
JP NC	C = 0	Jump if carry (flag) not set
JP P	S = 0	Jump if sign (flag) positive (0)
JP M	S = 1	Jump if sign (flag) negative (1)

on the state of the zero, carry and sign flags, respectively. Hence a typical instruction is

```
JP   NZ,SUM
```

The effect of this instruction is to jump or branch out of sequence to the instruction with the symbolic address label SUM only if the zero flag is not set (Z = 0). Otherwise continue in sequence (Z = 1).

To illustrate the use of this instruction, consider the example introduced earlier, i.e. a program segment to sum together one hundred numbers which are stored in memory starting at address 2800 (hex). As was indicated, it is necessary to use a conditional jump instruction to control the branching or looping operation so that it ceases after the one hundred values have been summed together.

Program example 5.1: conditional jump instructions

```
         .
         .
         LD   B,00      Initialise sum to zero
         LD   HL,2800   Initialise HL to point to
                           first value
LOOP:    LD   A,B       Add first (next) value to
         ADD  A,(HL)       running total
         LD   B,A       Save running total (sum)
                           in B
         INC  HL        Increment HL to point to
                           next value
         LD   A,L       Have 100 values been added?
         CP   64
         JP   NZ,LOOP   No (Z=0): loop back to add
                           next value
                        Yes (Z=1): continue in
                           sequence (sum in B)
```

The program segment shown above sums together one hundred numbers which are stored in one hundred consecutive memory locations starting at address 2800 (hex). The following registers are used:

B   is used to hold the running total and the final sum (assumed in the range of a single byte)

H,L is used as the memory pointer to each value in the list.

Q5.2–5.4

Since the first value is stored in memory address 2800 (hex) the last (100th) value will be stored in memory address 2863 (hex). Hence the loop count is automatically maintained in register L and therefore the compare immediate instruction is used to determine when the loop count reaches 64 (hex), i.e. when the contents of register L reach 64 (hex) the program will have looped 99 times and hence the 100 values will have been summed together.

5.4 Program design

Given an application task, before any actual program code is written it is first necessary for the programmer to produce a design for the program. This is an important and essential step in the production of an application program since without a suitable design it is usually impossible to plan the required logical sequence of actions required in order to perform the desired task. This is especially true when the resulting program contains a number of conditional and unconditional jump instructions.

A number of program design methods are now in operation and, in general, the method adopted is influenced by the eventual

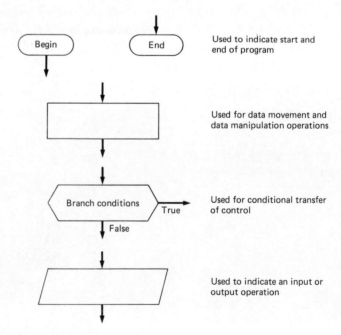

Figure 5.1 Flowchart symbols

programming language to be used. Since we are considering symbolic (assembly) languages in this book, however, perhaps the most suitable method is the *flowchart method* and hence this has been adopted throughout the remainder of this book.

Using the flowchart method the design for the program is expressed in a graphical or pictorial form. This is made up of a number of different shaped boxes or symbols each indicating a different type of operation and containing descriptive information describing the specific operations to be performed. These are then interconnected by unidirectional lines which are used to indicate the direction of flow and hence the logical sequence the operations are to be executed.

The symbols available are, in fact, closely related to the different instruction types available with a microprocessor instruction set. There is therefore a symbol to indicate data movement and data manipulation operations, a symbol to indicate conditional transfer of control operations and a symbol to indicate I/O operations. These are illustrated in Figure 5.1.

To illustrate the use of the two symbols concerned with data movement/arithmetic and conditional transfer of control, a design for the program segment shown earlier in Program example 5.1 is shown in Figure 5.2. It can be deduced from this figure that the flowchart design for a program is often at a slightly higher level than the

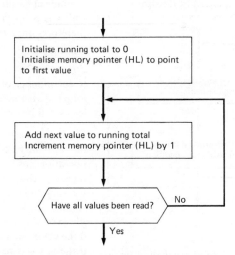

Figure 5.2   Flowchart for program example 5.5

final program. Thus the conditional symbol 'Have all the values been read?' does in fact require three instructions to implement. Indeed, with more sophisticated problems the design procedure itself is an iterative process that starts with a system-level flowchart and only after several steps of refinement is the final flowchart suitable for implementation in actual program code evolved.

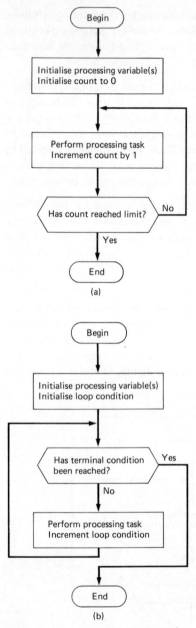

Figure 5.3   Two alternative looping flowcharts

Q5.5

Looping

Program example 5.1 is an example of the transfer of control construct which is known as 'looping with a known count'. This is a very common procedure and is sometimes used several times over within a single program.

The flowchart illustrated in Figure 5.2, however, is only one solution to this type of problem and sometimes an alternative flowchart is used. The two alternatives are shown in Figure 5.3. It can be deduced from the second solution that the conditional test operation is done before the processing task in the loop is performed. This can sometimes be necessary if, for example, the processing task in the loop should not be performed at all.

To show an application which uses solution *(b)*, consider the following program example. The program can be used to compute a variable time delay which is proportional to a value loaded into one of the microprocessor registers. A program has frequently to compute a time delay to ensure, for example, a specific time delay between outputting two consecutive values to an I/O interface port.

Program example 5.2:   looping with a known count

The flowchart and associated program code shown in Figure 5.4 is an example of looping with a known count. The program computes a time delay which is proportional to the value loaded into register C. The value can therefore range from 00 to FF (hex).

Since the microprocessor takes a small but finite time to execute an instruction (typically a few microseconds), the computed delay is proportional both to the number of instructions in the loop and also the number of times the loop is performed. Hence the larger the value loaded into register C, the more loops will be performed and hence the computed time delay will be longer.

The basic time delay instructions used within the loop are three no-operation or NOP instructions. These, as the name implies, perform no specific operation in so far as they affect none of the contents of the microprocessor registers. But, since the microprocessor must fetch the instruction from memory and decode it, the microprocessor automatically takes a few microseconds to fetch and execute each NOP instruction. The basic time delay (three NOPs) is therefore in the order of 10 µs. Notice that with this solution, if the value loaded into register C is zero, then no delay is computed, whereas if the test was performed at the end of the program after the delay instructions, at least one delay would be performed.

Binary decisions

Another very frequently required programming task is to perform one of two alternative processing operations depending on the result of a specified condition, i.e. IF the condition is met (true) THEN

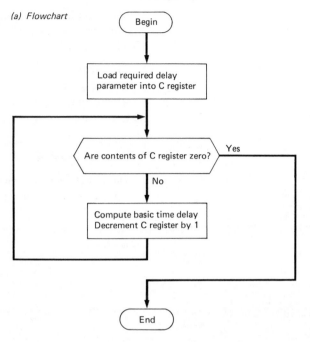

(a) Flowchart

(b) Program code

Symbolic instructions		Action
LD	C,XX	Load required delay parameter into C
DELAY: LD	A,C	Are contents of C zero?
CP	00	
JP	Z,TIME	
NOP		No. Perform basic time delay
NOP		
NOP		
DEC	C	Decrement delay parameter and loop back
JP	DELAY	
TIME:		Yes. Time delay computed

Figure 5.4   Program example 5.2

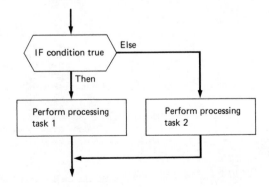

Figure 5.5   The binary decision or IF THEN ELSE construct

perform one processing operation ELSE perform another operation. This is known as the *binary decision* construct or alternatively, the IF THEN ELSE construct. It is shown in flowchart form in Figure 5.5.

To illustrate an application of this construct and how it is implemented in program code consider the following program example which may typically be encountered in the design of a simple temperature control program.

Program example 5.3:   the binary decision construct

The flowchart and associated program code shown in Figure 5.6 are an example of a program segment which uses the binary decision or IF THEN

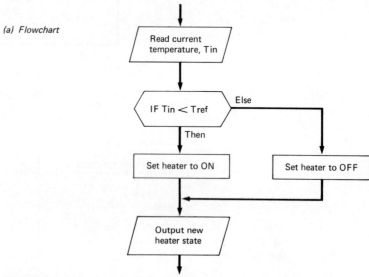

(a) Flowchart

(b) Program code

Symbolic instructions	Action
⋮	B holds Tref
⋮	C holds current heater state
Input Tin to register A	
CP B	IF Tin < Tref
JP NC,HOFF	
LD A,C	THEN set heater to ON
OR 01	
LD C,A	
JP HCON	
HOFF: LD A,C	ELSE set heater to OFF
AND FE	
LD C,A	
HCON: Output new heater	
state (C)	
⋮	
⋮	

Figure 5.6   Program example 5.3

ELSE construct. The program could be used to determine if the heating element used in a simple temperature control system should be switched on or off. This would typically be decided by the state of the current temperature in relation to the desired (set point) temperature: IF the current temperature (Tin) is less than the set temperature (Tref) THEN switch or leave the heater switched on ELSE switch the heater off. The heater can therefore be considered as a two-state device and hence controlled by a single bit.

The following microprocessor registers are used:

A holds the current temperature, Tin (typically read from an I/O input port).

B holds the set (desired) temperature, Tref.

C the least significant bit of C holds the current state of the heater (0 = off, 1 = on). (Typically other bits would control other devices.)

The program determines if the current temperature is less than the set temperature by means of the compare instruction. If after the compare instruction the carry flag is set, then the decision true (Tin < Tref) and hence the heater is either switched or left on else the heater is switched off. The logical OR and AND instructions are used to set and reset the heater bit, respectively, and hence the manipulation of this bit will not affect the current state of any of the other bits.

Use of subroutines

When designing a program it is quite common to require a particular operation (and hence piece of program code) several times over in the same program. For example, assuming the microprocessor being used does not contain a multiply instruction, it may be necessary to use the piece of code (list of instructions) which performs the multiplication operation several times within the same program. Clearly, each time the code is executed the only difference is likely to be the values on which the code operates rather than the instructions themselves.

This can therefore be a tedious and potentially inefficient use of memory. Consequently, all microprocessors include a pair of instructions which allow the programmer to write an often-used piece of code once only and then branch to it from within a program whenever the particular operation is required.

Since the values to be processed by the code are likely to differ each time the operation is carried out, it is usual for the programmer to arrange that the values to be processed are placed (stored) in a known area of memory prior to the branch taking place. Similarly, any results produced by the code are also placed in a known area and hence when the (main) program is re-run, it can readily access the results.

The piece of code that performs the often-used operation is called a *subroutine* and the instruction that causes the branch to it a *CALL*

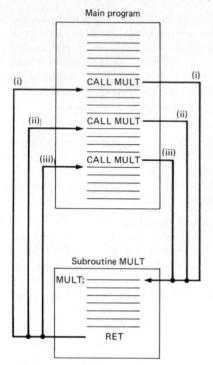

Figure 5.7   Subroutine CALL and RETurn mechanism

instruction. Clearly, after the subroutine has run it is necessary to branch back or return to the point in the main program from where it was called. This is achieved using the *RETurn* instruction and Figure 5.7 illustrates how a single subroutine may be called a number of times from different points within the same program.

In addition to avoiding a lot of repetitive code when writing a program (and hence saving memory), subroutines play an important role in the design of a well structured and, hence, more easily read and understood program. The facility of being able to write a separate piece of code to perform a specific operation and the use of a well defined mechanism for communicating with it, provides the programmer with an ideal vehicle for breaking down what would perhaps be a long complex program into a number of smaller, more easily understood sub-programs (subroutines). Moreover, it then becomes possible to utilise a number of programmers when designing and writing a large program since each can design a number of the smaller sub-programs.

The structure of a typical application program containing a number of subroutines is as shown in Figure 5.8. As will be seen in the next section, it is quite acceptable to call another subroutine from within a subroutine and hence the main program segment can often be just a small piece of code which initiates calls to the subroutines in the appropriate sequence.

Although the programs in this book will in general be quite small, the use of subroutines in program design is very important and hence subroutines will be used wherever possible to illustrate the technique involved. The next section describes the implementation of subroutines in more detail.

5.5   Subroutine implementation

It can be concluded from Figure 5.7 that the CALL instruction is very similar to an unconditional jump instruction: the first instruction in the subroutine is given a symbolic address name or label and this is used each time the subroutine is called. The difference can perhaps best be seen not by considering the calling mechanism but rather by considering the implications of the RETurn instruction.

Although a branch occurs to the same place each time the CALL instruction is executed, the branch to be implemented when the RET instruction is executed is determined by the point in the main program (and hence the CALL) from which the subroutine was first run. Notice in the figure that the branch back to the main program always takes place to the next sequential instruction after the CALL which initiated the running of the subroutine.

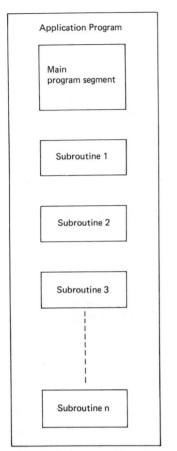

Figure 5.8 Application program structure using subroutines

Hence the destination address to be used when the RET instruction is executed is the next sequential address after the corresponding CALL instruction. This is, of course, the contents of the microprocessor program counter (PC) prior to the branch to the subroutine being carried out. When executing the CALL instruction, therefore, the microprocessor saves the current contents of the program counter so that when the corresponding RET instruction is executed, it knows where in the calling program to branch back to. It simply restores the saved address into the program counter and hence automatically branches back to the correct place within the calling program.

The stack

As has been mentioned, it is quite acceptable for the programmer to CALL another subroutine from within a subroutine. This is shown diagrammatically in Figure 5.9 and is known as *nesting subroutines*.

The implication of this mechanism is that the microprocessor may have to save a number of return addresses before executing a RETurn instruction. For example, when executing the program shown in Figure 5.9, the microprocessor would have to save one return address when CALL SUB1 was executed and another when CALL SUB2 was executed. Moreover, these saved addresses are required in the reverse sequence since the RETurn instruction at the end of subroutine 2 will be executed before that in subroutine 1.

The mechanism employed by a microprocessor to store or remember the return address associated with subroutine calls is known as a *stack*. A stack is simply a last-in, first-out queue since all the transactions take place at the head of the queue. Thus, when a value is entered into the queue, it is placed on top of those already in the queue and when a value is removed from the queue it is always the value currently at the head of the queue.

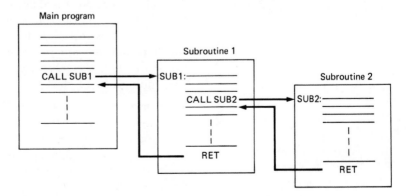

Figure 5.9 Diagrammatic representation of nested subroutines

This is therefore an ideal mechanism for handling subroutine return addresses since each time a CALL instruction is executed the next sequential address (the return address) is placed on top of the stack (queue) so that when the RETurn instructions are executed, the corresponding return addresses will always be at the head of the queue.

Entering a value into a stack is known as *PUSHing* the value and removing a value from a stack *POPing* the value. Thus when the microprocessor executes a CALL instruction, it saves the current contents of the program counter (the return address) by PUSHing it onto the stack and then branches to the start of the subroutine. Similarly, when the microprocessor executes a RETurn instruction, it first POPs the return address from the head of the stack and then loads it into the program counter.

The stack is normally implemented in a microcomputer system as part of the normal memory space. The microprocessor thus contains a special 16-bit register called the *stack pointer* which always holds the address in memory of the top of the stack. It can then enter and remove values (return addresses) to and from the stack very readily simply by accessing and manipulating its stack pointer.

Subroutine parameters

One of the primary reasons for using a subroutine is that the code required to implement an often-used processing task need be written just once only and hence the code does not need to be replicated. The values to be processed by a subroutine, however, may clearly vary each time the subroutine is called. For example, if we write a subroutine to multiply two values together, then the code will be the same but the two values to be multiplied are likely to be different each time the subroutine is executed.

The values to be processed by a subroutine are known as *parameters* (or *arguments*) and these are said to be *passed* to the subroutine for processing. It is therefore necessary to establish a mechanism by which parameters are passed between a calling program and a subroutine and *vice versa* for any results that may be produced by the subroutine – the product in the above example.

The mechanism used varies, but perhaps the most common approach is to use an area within the memory. The calling program then places the value(s) to be processed in a known order in this block of memory (list of locations) and then passes the subroutine the start address of the block. The subroutine in turn accesses the value(s) to be processed from the block (in the correct sequence), processes them and then places the result(s) (also in a known sequence) into the block. The calling program can then access the result(s) from the block and continue. This approach will be used in the following program examples which have been written to illustrate some of the points just discussed.

Program example 5.4: *parameter passing*

```
        .
        .
        .
        .
        LD      HL,2050     Initialise HL to point to memory block
        LD      (HL),00     Store start address of list of instructions, 2800
        INC     HL          (hex), in memory block
        LD      (HL),28
        DEC     HL          Restore HL to point to start of memory block
        CALL    SUM         Call subroutine SUM
        LD      B,(HL)      Load sum into B and continue
        .
        .
        .
        HALT                End of main program
------------------------------------------------------------------------
SUM:    LD      E,(HL)      Load start address of list of instructions in DE
        INC     HL
        LD      D,(HL)
        INC     HL          Obtain start address in HL and save memory block
        EX      DE,HL       pointer in DE
        LD      B,00        Initialise sum to zero
LOOP:   LD      A,B         Add first (next) value to running total
        ADD     A,(HL)
        LD      B,A         Save running total (sum) in B
        INC     HL          Increment HL to point to next value
        LD      A,L         Have 100 values been added?
        CP      64
        JP      NZ,LOOP     No: loop back to add next value
        EX      DE,HL       Yes: obtain memory block pointer from DE
        LD      (HL),B      Store sum in memory block
        RET                 Return to calling program
```

The program segment shown above illustrates the mechanism of passing parameters to a subroutine by means of a memory block. The subroutine SUM computes the sum of a list of 100 values held in memory similar to the program segment shown earlier in Program example 5.1. The start address of the list of values and the computed sum (assumed 8-bits) to be returned from the subroutine are stored in a block of memory starting at address 2050 (hex) as follows:

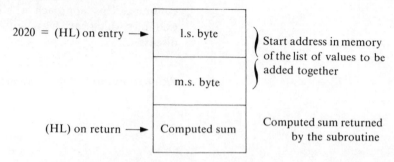

2020 = (HL) on entry ⟶ | l.s. byte

m.s. byte

} Start address in memory
of the list of values to be
added together

(HL) on return ⟶ | Computed sum   Computed sum returned
by the subroutine

The microprocessor registers are used in the subroutine in the same way as
they were used earlier in Program example 5.1. Hence it is assumed that
register L can also be used as a counter to determine when all 100 values have
been added together. A different list of values could readily be added together
by the subroutine simply by passing the start address of the new list in the
memory block. To continue using L as a counter, however, the least
significant address byte must always be 00 (hex).

Q5.8–5.10

Program example 5.5:   nested subroutines

```
        .
        .
        .
        LD      HL,2800   Load delay parameter (XX) into memory location
        LD      (HL),XX   2800 (hex)
        CALL    DELAY1    Call first subroutine
        .
        .
        .
```
- -
```
DELAY1  LD      C,(HL)    Load delay parameter into C
DELAY:  LD      A,C       Are contents of C zero?
        CP      00
        JP      Z,TIME
        INC     HL        No: load second delay parameter (YY) into memory
        LD      (HL),YY   location 2801 (hex) and call second subroutine
        CALL    DELAY2
        DEC     C         Decrement delay parameter and loop back
        JP      DELAY
TIME:   RET               Yes: return to calling program
```
- -
```
DELAY2: LD      B,(HL)    Load delay parameter into B
LOOP:   LD      A,B       Are contents of B zero?
        CP      00
        JP      Z,END
```

(Continued)

```
        NOP                 No: perform basic time delay
        NOP
        NOP
        DEC    B            Decrement delay parameter and loop back
        JP     LOOP
END:    RET                 Yes: return to calling program
```

The program segment shown above illustrates the ability for the programmer to call another subroutine from within a subroutine. Although it is possible to extend the time delay computed by the program shown earlier in Figure 5.4 by replacing the three NOP instructions by, say, a simple looping operation with a known count, the delay has been extended here by replacing the NOP instructions by another CALL instruction to a further delay subroutine. The computer time delay is therefore extended by calling the second sub-routine each time a loop occurs in the first subroutine.

The delay parameter for the first subroutine (DELAY1) is passed in memory location 2800 (hex) and the delay parameter for the second subroutine (DELAY2) in location 2801 (hex).

Stack operations

Program example 5.5 highlights an important point about the use of subroutines. It can be seen from the program listing in that example that the second subroutine (DELAY2) uses a different register (register B) to hold the delay parameter from that used by the first subroutine (DELAY1). This is because the second subroutine would otherwise make the contents of register C zero when it was run so that on return, the first subroutine would prematurely return to the main program after just a single loop.

In this instance it was possible to avoid this happening by the use of a second register but usually this is not possible because both the main program and the subroutine(s) often use all the registers available. It is for this reason that when a subroutine is written, it is common practice to save the current contents of any microprocessor registers used within the subroutine immediately on entering the subroutine and to restore them to their original state prior to returning to the calling program. Each program segment can then freely use all the available registers in the knowledge that they will not be corrupted or changed when a subroutine is called.

In addition to the microprocessor using an area of memory as a stack to hold subroutine return addresses, it is normal for a microprocessor to then provide instructions to allow the user to use the stack as a temporary repository for the contents of the various microprocessor registers. Thus in the Z80, the four instructions

```
PUSH   AF
PUSH   BC
PUSH   DE
PUSH   HL
```

are provided to store (PUSH) the current contents of the above list of register pairs on the stack. Similarly the following instructions are provided for reloading the various register pairs from the stack:

```
POP   AF
POP   BC
POP   DE
POP   HL
```

Since the stack operates in a last-in, first-out manner, it is essential

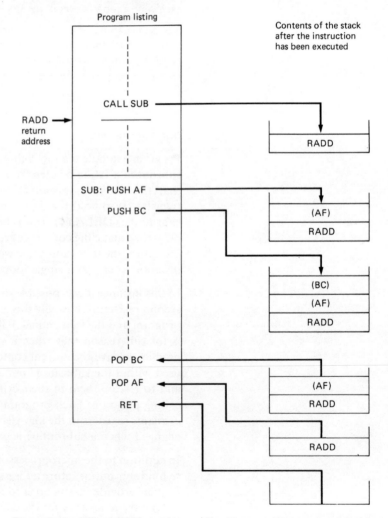

Figure 5.10   Effect of different instruction types on the contents of the stack

that the programmer orders the PUSH and POP instructions so that the correct contents are restored to the correct register pair prior to leaving the subroutine. For example, assuming a subroutine utilised registers A (and hence F) and C within it, it would be normal for the programmer to save the contents of both register pairs A,F and B,C at the start of the subroutine and then restore them at the end. This would therefore be done as shown in Figure 5.10.

It can be concluded from this figure that the POP instructions must always be in the reverse order to the corresponding PUSH instructions. Also, since the microprocessor itself uses the stack to save the return address, it is essential to ensure that the same number of POP instructions are used at the end of the subroutine as PUSH instructions at the start. Hence when the microprocessor executes the RET instruction, it will restore the correct return address into the program counter from the stack.

Q5.11

The contents of the stack have been shown for convenience in the figure as single 16-bit words but in practice these will be stored as two consecutive bytes in memory.

Questions

5.1 Using the information contained in Appendix I, translate the following program segment into its hexadecimal machine code form assuming the first byte of the first instruction is stored at memory address 2000 (hex):

```
LOOP:   LD    A,E7
        INC   A
        JP    LOOP
        -
```

5.2 Determine the final contents of register B after the following program segment has run:

```
        LD    B,FF
NZERO:  DEC   B
        JP    NZ,NZERO
        -
```

5.3 Translate the program segment given in Question 5.2 into its hexadecimal machine-code form, assuming the first byte of the first instruction is stored at memory address 2020 (hex).

5.4 Determine the final contents of register A after the following program segment has run and also the number of instructions that will have been executed:

```
            LD    A,00
    LOOP:   INC   A
            JP    NC,LOOP
            -
```

5.5   Design a flowchart and write a symbolic (assembly) language program to add together the numbers 1, 2, 3, ..., 10. (*Hint:* Use one of the microprocessor registers to hold the running total and another to hold the next number to be added to the running total. The program can then be conveniently structured using a loop with a known count.)

5.6   Design a flowchart and write a symbolic language program segment which either adds 15 (dec) to the contents of register A if the current contents of register A are zero or adds 20 (dec) to the contents of register A if the current contents of register A are non-zero. The contents of the register A are assumed unknown when the program segment is executed.

5.7   Repeat Question 5.6 with the condition to be satisfied to add 15(dec) to the contents of register A modified to be less than or equal to zero, and the condition to add 20(dec) greater than zero.

5.8   Modify the program developed in Question 5.5 into a subroutine with the number range and the computed total to be passed as parameters in a memory block whose start address is passed in register pair H,L.

5.9   Modify Program example 5.2 so that the three basic time delay instructions (NOPs) are replaced by a simple looping operation. The loop should be arranged to be a loop with a known count of 100.

5.10 If each instruction takes 4 µs to fetch and execute, determine the minimum and maximum time delay computed by the program developed in Question 5.9.

5.11 Modify the subroutine developed in Question 5.8 so that the parameters are passed by means of the stack, i.e. the range of the numbers to be added together is PUSHed onto the stack prior to calling the subroutine and the result is PUSHed onto the stack prior to returning. (*Hint:* remember the return address will be at the top of the stack when the subroutine is first executed and hence must be POPped and saved whilst the computation is performed.)

Chapter 6 I/O instructions and associated programming techniques

Objectives of this chapter *After you have completed studying this chapter you should be able to:*

1 *Understand and use the basic input and output instructions provided with a microcomputer.*
2 *Describe typical devices which produce or use digital signals and be able to write short programs to interface a microcomputer to them.*
3 *Describe the function of a decimal and hexadecimal keypad interface circuit and be able to write a program to enter a string of digits from such keypads.*
4 *Explain the alternative display devices used with microcomputers and write programs to demonstrate their use.*
5 *Understand how data can be input and output by a microcomputer in a bit serial form.*
6 *Explain the term asynchronous transmission and the functions performed by a UART.*
7 *Describe the principle of operation of a DAC and how an analogue value can be generated by a microcomputer.*
8 *Describe the principle of operation of an ADC and how an analogue voltage can be monitored by a microcomputer.*

6.1 Introduction

A microcomputer can be considered as a device that simply reads digital input data, processes this data according to the list of instructions (program) stored within its memory and then outputs the subsequent results in the form of digital output data. Moreover, although the actual input and output devices used will vary depending on the application, this is in many ways transparent to the microcomputer since suitable interface circuits are used so that a standard digital I/O mechanism can be employed which is independent of the application. Thus a digital (binary) value input or output using the appropriate microprocessor instructions may be from or used by a variety of different devices depending on the application and the particular interface circuits employed.

This chapter is concerned with the basic input and output instructions normally provided with a microprocessor and, in addition, describes both typical I/O devices used with microcomputers and the associated programming techniques required to use them.

6.2 I/O ports

The I/O instructions provided with a microprocessor simply input or output an 8-bit binary coded value to or from a specified port within the I/O interface unit. In its simplest form, therefore, the I/O interface unit can be considered as a collection of input and output ports connected to the microcomputer bus as shown in Figure 6.1.

The number of ports available can normally be very large – typically 256 (decimal). Hence, although for many applications the actual number of ports used is often quite small, the input and output instructions available with a microprocessor normally specify a port (input or output) within the range 00 to FF (hex). Thus a typical input instruction provided by the Z80, for example, is

    IN A,(01)

This has the effect that the port address [01 (hex) in this example] is placed on the address lines of the microcomputer bus by the microprocessor and the I/O interface unit responds by enabling the current 8-bit value which is present at the input of the addressed port onto the data lines of the bus. This is then loaded into register A by the microprocessor. Thus if the current value present at the input of the port was, say, E7 (hex), then after the above instruction had been executed, the contents of register A would be E7 (hex).

Similarly, a typical output instruction provided by the Z80 is:

    OUT (02),A

This causes the microprocessor to output the current contents of register A onto the data lines of the microcomputer bus and the port address [02 (hex)] on the address lines of the bus. The I/O interface unit then loads or latches this value into the selected port buffer register. Thus, if the contents of register A were, say, 7F (hex) prior to execution of the above instruction, this value would be loaded into port register 02 (hex) after the instruction had been executed.

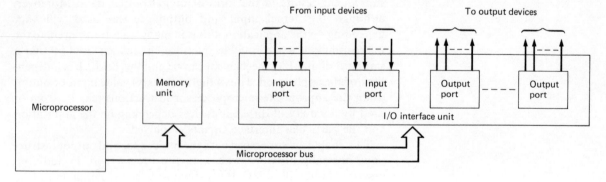

Figure 6.1 I/O ports of a microcomputer

The input and output ports serve two important functions: on input, they are used to isolate any data values produced by an input device or interface circuit from the microcomputer bus until the microprocessor executes an INput instruction on this port and on output, they are used to hold or latch a data value output by the microprocessor until the output device or interface circuit is ready to accept it. The input and output ports are therefore also referred to as I/O buffers since they form a buffer between the microcomputer bus and the I/O devices and interface circuits being used for the application.

6.3 Digital input and output

Since the data input and output at the I/O interface ports are digital in nature (two state), it is possible to readily interface digital devices to a microcomputer. Thus the state of a row of up to eight switches (on–off) or push-buttons (open–closed) can be input directly from an input port. Similarly, a row of up to eight light-emitting diodes (LEDs) can be driven directly to display a binary-coded value output from an output port.

Alternatively, with suitable interface circuits connected to the ports, a binary value entered at an input port may indicate the current state of, say, a set of level switches or thermostats connected to a controlled process. Similarly, a binary value output to a port may be used to operate a set of relays or solenoid valves connected to the process. Some alternative I/O devices and the associated programming techniques used to control them are now described.

Digital input and output from a set of switches and LEDs

The simplest form of input and output devices are a set of switches to

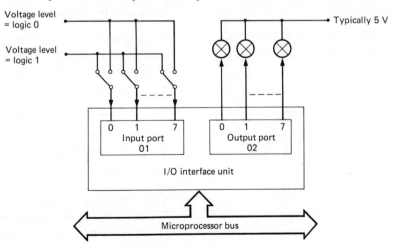

Figure 6.2 Digital I/O using switches and LEDs

enter a binary value directly and a set of LEDs to display a binary value output by the microcomputer. Normally, both these devices can be connected directly to an input and output port respectively without the use of additional interface circuitry. A typical arrangement is shown in Figure 6.2 and some program examples which use this circuit are now given.

Program example 6.1

The flowchart and associated program code shown in Figure 6.3 is for a program that continuously reads the state of the eight switches and outputs

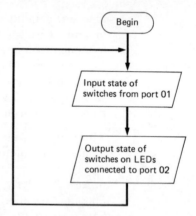

Symbolic instructions	Action
LOOP: IN A,(01)	Input state of switches from port 01 into A
OUT (02),A	Output contents of A to port 02
JP LOOP	Loop

Figure 6.3   Program example 6.1

the value read to the eight LEDs shown in Figure 6.2. Since the state of the switches are input within the program loop, any changes in the state of the switches will be immediately reflected in the state (on or off) of the corresponding LEDs.

Program example 6.2

The program described in Program example 6.1 will clearly loop forever simply reading the state of the switches and outputting the states on the LEDs. The program shown in Figure 6.4, however, reads the state of the switches and displays their states on the LEDs until a specific pattern [AA (hex) in this example] is read from the switches. The program will then terminate.

Q6.1

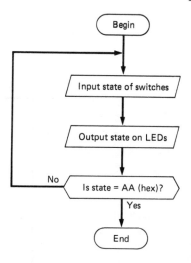

Symbolic instructions	Action

LOOP: IN   A,(01)
        OUT (02),A
        CP   AA
        JP   NZ,LOOP
        HALT
	Input state of switches into A
	Output contents of A to port 02
	Is state = AA (hex)?
	No: loop
	Yes: end

Figure 6.4   Program example 6.2

Program example 6.3

The program of Figure 6.5 illustrates how a computed time delay may be used within a program. The program simply counts from 00 to FF (hex) and displays within the counting loop the current count on the LEDs. Since the basic counting loop is very fast, however, the least significant bits of the display would change too quickly to follow. The switches are used in this example, therefore, to enter an 8-bit value which is then used as a parameter to determine the delay computed between displaying successive values in the count. Thus the larger the value set on the switches the slower the count will change.

The computer delay routine is similar to that shown earlier in Figure 5.4. The registers used by the program are A (general working register), B (holding the current count) and C (holding the delay parameter).

Q6.2

Digital input and outut from a controlled process

With suitable circuits to provide an electrical interface between the input and output ports of the microcomputer, the digital inputs considered previously, instead of being derived from a set of switches could readily be derived from various digital devices associated with, say, a process which is being controlled by the microcomputer. For

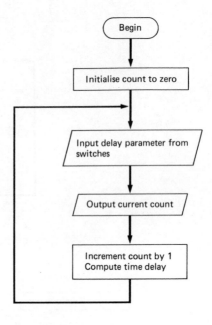

Figure 6.5 shows the flowchart:
- Begin
- Initialise count to zero
- Input delay parameter from switches
- Output current count
- Increment count by 1 / Compute time delay

Symbolic instructions			Action
	LD	B,00	Initialise count to zero
LOOP:	IN	A,(01)	Input delay parameter from switches
	LD	C,A	Store delay parameter in C
	LD	A,B	Output current count
	OUT	(02),A	
	INC	B	Increment count by 1
	LD	A,C	Compute time delay
DELAY:	CP	00	
	JP	Z,LOOP	
	NOP		
	NOP		
	NOP		
	DEC	A	
	JP	DELAY	

Figure 6.5   Program example 6.3

example, a set of thermostats (open-closed) or limit switches (on-off). Similarly, the digital outputs could be used to control the state of various solenoid valves (open-closed) and relays (on-off) associated with the process. This is shown diagrammatically in Figure 6.6.

The input and output programming techniques performed on this data are clearly different from what has just been considered. Some examples of typical processing actions associated with digital inputs and outputs from a controlled process are now considered.

Program example 6.4:   sequencing I

A common processing action when controlling a process is to output a sequence of states which are typically stored within a table in the micro-

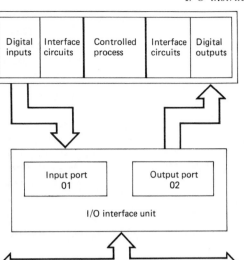

Figure 6.6 Digital input and output from a controlled process

(a) State table

State number	Controlled device states							
	8	7	6	5	4	3	2	1
0	0	0	0	1	0	1	0	1
1	0	0	1	0	1	0	1	1
2	0	0	1	1	1	1	1	1
3	0	1	0	0	1	0	0	1
4	0	1	0	1	0	1	0	0
5	1	1	0	0	0	0	1	0
6	0	0	0	0	0	0	0	0

(b) Flowchart and program code

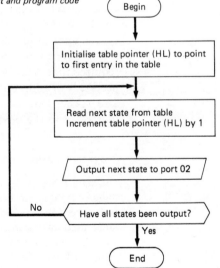

Symbolic instructions			Action
	LD	HL,2800	Initialise H,L to first entry in table
REPEAT:	LD	A,(HL)	Read next state from table
	OUT	(02),A	Output next state to port 02
	INC	HL	Increment table pointer by 1
	LD	A,L	Have all states been output?
	CP	07	
	JP	NZ REPEAT	No: repeat
	HALT		Yes: end

Figure 6.7 Program example 6.4

computer memory. For example, each stored value within the table may indicate the required state (on or off) of a set of valve solenoids or relays which are connected to the microcomputer output port via suitable drive circuits. This control function is therefore known as *sequencing* and the microcomputer as the *sequence controller*.

An example of a typical sequence of actions of eight controlled devices may therefore be as illustrated in Figure 6.7(a). This is normally referred to as a sequence or *state table* for the system. Thus the first entry in the table indicates that devices 1, 3 and 5 should be turned on and that devices 2, 4, 6, 7 and 8 should be turned off. Similarly, the next entry in the table indicates the next state to be output and shows that devices 1, 2, 4 and 6 should now be turned on and that devices 3, 5, 7 and 8 turned off. This then continues until all the devices are turned off at the end of the sequence – (state 7).

A flowchart and the associated program code to perform this sequencing action is therefore as shown in Figure 6.7(b). The program simply reads each new state from the table [which is assumed to be stored in memory starting at address 2800 (hex)] and then outputs this state (value) to the output port. This is then repeated until all seven states have been output.

Q6.3

The program shown in Figure 6.7 simply performs a looping operation to access each new set of device states from the table and then outputs this to the appropriate output port. The time delay between each new set of device states being output will therefore be very short and for many applications this will clearly be insufficient. With this type of application, therefore, it is more usual for the microcomputer to compute a time delay between outputting each new state. Since the required time delay between each state may be different, however, it is necessary in addition to specifying the required set of device states, to also specify the time delay required between each state. This is illustrated in the next example.

Program example 6.5:   sequencing II

The state table shown in Figure 6.8(a) indicates the required sequence of states of eight controlled devices and the corresponding time delay between each state. The flowchart and program code shown in Figure 6.8(b) therefore outputs each state and then uses the associated time delay specified in the table as a parameter for a time delay subroutine similar to the one shown earlier in Figure 5.4. The basic delay used in the subroutine must be made equal to one second and hence the delay parameter indicates the number of loops to be performed within the subroutine. Since the times indicated are in decimal, these must first be converted into hexadecimal form before storing in a table.

The program assumes the state table data are stored within a table in memory starting at address 2800 (hex). For convenience, a single table is used in which each set of device states is immediately followed by the required time delay parameter. Thus the first value in the table (at address 2800) will be 15 (hex) to indicate the required device states and the next value (at address 2801) will

(a) State table

State number	Time delay between states, s	Controlled device states							
		8	7	6	5	4	3	2	1
0	1	0	0	0	1	0	1	0	1
1	20	0	0	1	0	1	0	1	1
2	120	0	0	1	1	1	1	1	1
3	120	0	1	0	0	1	0	0	1
4	1	0	1	0	1	0	1	0	0
5	250	1	1	0	0	0	0	1	0
6	0	0	0	0	0	0	0	0	0

(b) Flowchart and program code

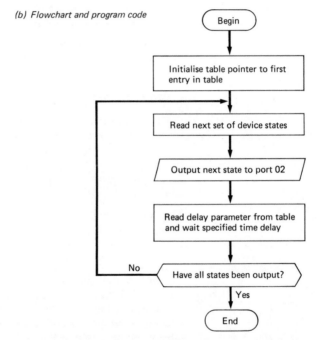

Symbolic instructions			Action
	LD	HL,2800	Initialise table pointer
REPEAT:	LD	A,(HL)	Read next set of device states from table
	INC	HL	
	OUT	(02),A	Output next state to port 02
	LD	C,(HL)	Read delay parameter from table and wait
	CALL	DELAY	specified time delay
	INC	HL	Have all states been output?
	LD	A,L	
	CP	0E	
	JP	NZ REPEAT	No: repeat
	HALT		Yes: end
DELAY:	–		Time delay subroutine similar to that
	–		in Figure 5.4
	.		
	.		
	.		
	RET		

Figure 6.8 Program example 6.5

Q6.4

be 01 (hex) to indicate the required time delay. The final set of states and associated time delay value will therefore be stored at memory addresses 280C (hex) and 280D (hex), respectively, and hence the loop operation terminates when register L reaches 0E (hex).

The above two examples illustrate typical programming actions used to perform basic or *unconditional* sequencing. In some sequencing applications, however, instead of waiting for a specified time delay between outputting each new set of device states, it is necessary to wait for a certain set of input conditions (device states) to occur.

For example, if the set of output states control the operation of a set of valve solenoids which in turn control the flow of a fluid into several tanks, then it may be necessary to wait between each new output state for the state of an associated set of level switches to reach a certain condition. This mode of operation is then known as *conditional sequencing* and an example is given next.

Program example 6.6: *conditional sequencing*

The table shown in Figure 6.9(a) lists the conditional states required of a set of eight input devices after each new state of a set of eight controlled devices has been output. Thus after outputting the first set of device states, the microprocessor must wait until input devices 3, 4 and 6 are on and input devices 1, 2, 5, 7 and 8 are off before proceeding to output the second set of device states and so on.

The flowchart and program code of a program to perform this function is therefore as shown in Figure 6.9(b). As can be seen, after outputting each new state the program continuously loops waiting for the specified input condition to be satisfied.

The program assumes the output states are stored in a table in memory starting at address 2800 (hex). Again, for program efficiency, each set of conditional states of the eight input devices is stored in the table immediately after each corresponding set of output device states. Thus the first value in the table (at address 2800) will be 3C (hex) corresponding to the first set of output device states and the next value (at address 2801) will be 2C (hex) corresponding to the required conditional input states. Thus all states have been output when register L has been incremented to 0E (hex).

Q6.5

6.4 Some alternative input and display devices

It can be concluded from the previous section that for a number of microcomputer applications, the input and output of simple binary patterns or values is perfectly adequate. For others, however, more sophisticated facilities are required both for the source of the data to be processed and the display of the subsequent results. Some alternative devices and techniques are now considered.

(a) *State table*

State number	Conditional input states								Controlled device states							
	8	7	6	5	4	3	2	1	8	7	6	5	4	3	2	1
0	0	0	1	0	1	1	0	0	0	0	1	1	1	1	0	0
1	0	0	0	1	0	0	1	1	0	0	0	1	0	0	1	1
2	0	1	0	0	0	0	1	1	0	1	0	0	0	0	1	0
3	1	0	0	0	0	0	0	0	1	0	0	0	0	0	0	0
4	0	0	1	0	1	1	0	0	0	0	1	0	1	1	0	0
5	1	1	1	1	1	1	1	1	1	1	1	1	1	1	1	1
6	0	0	0	0	0	0	0	0	0	0	0	0	0	0	0	0

(b) *Flowchart and program code*

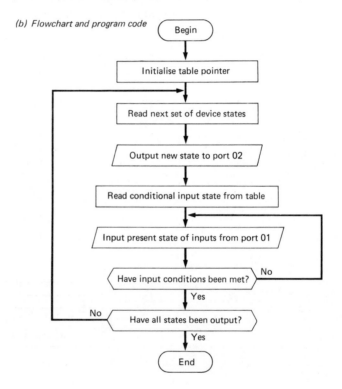

Symbolic instructions			Action
	LD	HL,2800	Initialise table pointer
REPEAT:	LD	A,(HL)	Read next set of device states from table
	INC	HL	
	OUT	(02),A	Output new state to port 02
	LD	B,(HL)	Read conditional input state from table
	INC	HL	
WAIT:	IN	A,(01)	Input present state of inputs from port 01
	CP	B	Have input conditions been met?
	JP	NZ,WAIT	No: wait until conditions met
	LD	A,L	Yes: have all states been output?
	CP	0E	
	JP	NZ,REPEAT	No: repeat
	HALT		Yes: end

Figure 6.9   Program example 6.6

Digital input from a keypad

Although a set of switches provides a simple means for entering a binary value into a microcomputer, for those applications which require, say, a number of decimal digits to be entered, it would be an inconvenient and time-consuming method. Consequently, since the latter is a common requirement in a number of applications, special interface circuits are available which perform the encoding process to convert the selection of a specific key on a keypad into a suitable binary-coded form which can be read by a microcomputer.

For example, a decimal or hexadecimal keypad encoder circuit produces the 4-bit binary-coded equivalent of a decimal or

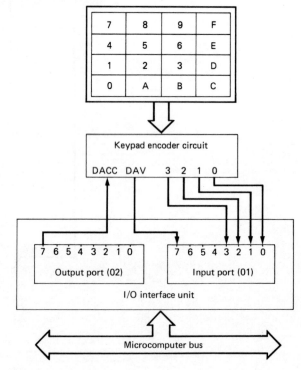

(a) *Keypad connection schematic*

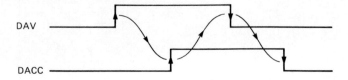

(b) *Handshake control*

Figure 6.10   Digital input from a keypad

hexadecimal digit selected on a ten- or sixteen-key keypad. In addition, since a string of digits may be entered at the keypad, it is also necessary to determine when a new digit has been selected so that the binary-encoded equivalent of the digit can be read by the microcomputer prior to the next digit being entered.

This is normally accomplished by the circuit generating a signal on a separate control line to indicate to the microcomputer when a new digit has been selected and in turn, the microcomputer generating a signal on another control line when it has read the digit. This procedure or *protocol* is known as a *handshake* procedure and is used extensively in digital systems to synchronise the transfer of data between two devices.

A schematic diagram illustrating how a hexadecimal keypad and its associated encoder circuit may be connected to a microcomputer using an input and an output port is shown in Figure 6.10(a). As can be seen, the handshake control lines are connected to bit 7 of the input port and also bit 7 of the output port. They are normally referred to as the data avaliable (DAV) and the data accepted (DACC) lines, respectively.

When a key has been depressed on the keypad, the keypad encoder circuit generates the equivalent 4-bit binary encoded digit on the 4 data lines and signals to the microcomputer on the DAV line that a new digit has been entered and is available for reading. The microcomputer then detects the setting of the DAV line, reads the digit and then responds by setting the DAC line to signal to the encoder circuit that the digit has been read. The latter than interprets this as an acknowledgement that the digit has been read and hence resets the DAV line. Finally, the microcomputer detects the resetting of the DAV line and in turn resets the DACC line ready for a new transfer.

Program example 6.7:   reading a digit from a keypad

The program segment shown in Figure 6.11 illustrates how a digit may be read from a hexadecimal keypad connected to the microcomputer I/O ports as shown earlier in Figure 6.10(a). The program first loops reading the value at the input port until the DAV line (bit 7) becomes set. It then accepts the digit and sets the DACC line (bit 7 of the output port) before looping once again waiting for DAV to be reset. Finally, when DAV is reset, the program resets the DACC line to complete the handshake cycle.

Q6.6

The previous program example illustrates how a single digit can be read from a keypad. It is often necessary, however, to read a string (number) of digits from the keypad prior to processing. This is readily accomplished by incorporating the code developed to read a single

(a) Flowchart

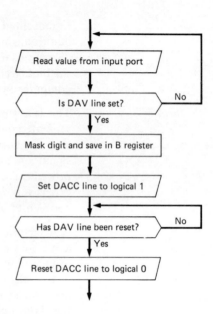

(b) Program code

Symbolic instructions			Action
INPUT:	IN	A,(01)	Read value from input port
	LD	B,A	Save in register B
	RLA		Is DAV line 1? (CY ← 1)
	JP	NC,INPUT	No: wait until DAV line set
	LD	A,B	Yes: mask digit and save in register B
	AND	0F	
	LD	B,A	
	LD	A,80	Set DACC line to 1
	OUT	(02),A	
WTDAV:	IN	A,(01)	Has DAV line been reset?
	RLA		
	JP	C,WTDAV	No: wait until DAV line reset
	LD	A,00	Yes: reset DACC line to 0
	OUT	(02),A	

Figure 6.11   program example 6.7

digit into a subroutine and then calling the subroutine to read each digit in the string. The final digit is then determined either by defining the number of digits in the string and employing a simple count mechanism or by defining a specific termination digit so that when it is read the string is complete and is now ready for further processing. This is illustrated by means of an example.

Program example 6.8:   *reading a string of digits from a keypad*

The flowchart and program code shown in Figure 6.12 uses the program segment derived in Figure 6.11 as a subroutine to enable a string of four hexa-decimal digits to be read from a keypad. The four digits may typically represent a 16-bit memory address and consequently the assembled 16 bits are stored in register pair H,L. The first digit entered is assumed to

(a) Flowchart

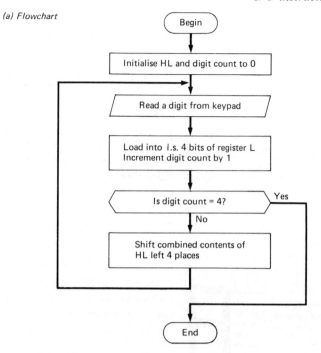

(b) Program code

Symbolic instructions			Action
	LD	HL,0000	Initialise H,L and digit count (C) to zero
	LD	C,00	
MORE:	CALL	INPUT	Read a digit from the keypad (left in B)
	LD	A,B	Load into l.s. 4 bits of register L
	ADD	L	
	LD	L,A	
	INC	C	Increment digit count
	LD	A,C	Is digit count = 4?
	CP	04	
	JP	Z,END	Yes: end (process string)
	SLA	L	No: shift combined contents of H,L left
	RL	H	four places
	SLA	L	
	RL	H	
	SLA	L	
	RL	H	
	SLA	L	
	RL	H	
	JP	MORE	Loop to read next digit
END:	HALT		End
INPUT:	-		Subroutine to read a digit from keypad
	.		
	.		
	.		
	RET		

Figure 6.12   Program example 6.8

represent the most significant digit and hence is stored in the most significant half of register H and so on.

The program utilises two shift instructions (SLA L and RL H) which are special to the Z80 for performing the combined left shift of the contents of register pair H,L. The first instruction (SLA L) shifts the contents of L left one place with the most significant bit being shifted into the carry bit (CY) and the least significant bit set to zero. The second instruction (RL H) is then used to shift the contents of H left one place with the least significant bit becoming equal to the current contents of the carry bit. Thus after repeating this pair of instructions four times, the combined H,L contents will be shifted left four places and four zero bits will occupy the least significant four bits of L.

Display of decimal digits

As has been described, a set of LEDs provides a simple means of displaying a binary value output by a microcomputer using an output port. Also, if each LED conveys a specific meaning (for example, a particular control valve is currently on or off), it provides a convenient method for displaying the status of a set of digital devices associated with a controlled process.

In a number of applications, however, a decimal display is required to display, for example, the measured weight in a microprocessor-controlled weighing machine, or the temperature of a controlled process on a control panel. It is clearly not acceptable to use four LEDs to display each digit in its binary-coded form and consequently a special interface circuit is normally used to perform the conversion process between the binary-coded outputs produced by the microcomputer and the decimal display.

The display employed for displaying decimal digits are usually seven-segment devices constructed using either LED segments or liquid crystal segments (LCDs). The display of each of the ten numeric digits on a seven-segment device is shown in Figure 6.13. The interface circuit used with these devices takes the 4-bit binary-coded value output by the microcomputer as input and produces the necessary drive signals to activate (on or off) the seven separate segments of the display.

Since it is often necessary to display a group of digits, interface circuits are available to enable up to four digits to be displayed from a single output port. A schematic diagram of such a circuit is shown in Figure 6.14. It can be deduced from this circuit that the four least significant lines from the output port are connected to all the latch/decoder/driver circuits and a particular digit position is selected using the four most significant lines. Thus the output of the binary value 0001 0111 from the output port would result in the decimal digit 7 being displayed at the least significant digit position. Similarly the

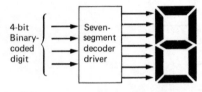

Decimal digit	Seven-segment representation
0	0
1	1
2	2
3	3
4	4
5	5
6	6
7	7
8	8
9	9

Figure 6.13 Seven-segment representation of a decimal digit

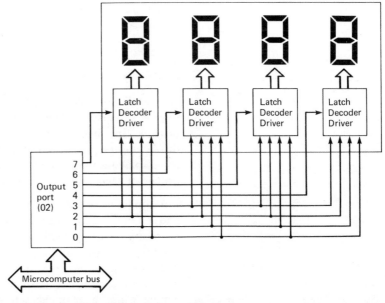

Figure 6.14 Four-digit decoder and display

binary value 1000 0110 would result in the decimal digit 6 being displayed at the most significant digit position and so on.

Program example 6.9: decimal display of numeric digits

The flowchart and associated program code shown in Figure 6.15 illustrates how four-binary coded digits held in register pair H,L are displayed using the

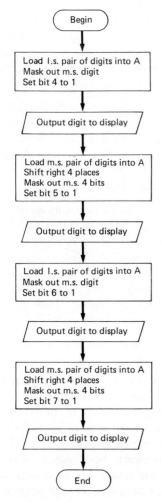

Figure 6.15 Program example 6.9

Symbolic instructions		Action
LD	A,L	Obtain first digit from L
AND	0F	
OR	10	
OUT	(02),A	Output to display
LD	A,L	Obtain second digit from L
RRA		
RRA		
RRA		
RRA		
AND	0F	
OR	20	
OUT	(02),A	Output to display
LD	A,H	Obtain first digit from H
AND	0F	
OR	40	
OUT	(02),A	Output to display
LD	A,H	Obtain second digit from H
RRA		
RRA		
RRA		
RRA		
AND	0F	
OR	80	
OUT	(02),A	Output to display

circuit shown in Figure 6.14. Each unwanted digit is first masked from the appropriate register (H or L) and then the appropriate selection digit added prior to it being output to the specified port.

Display of alphanumeric digits

A seven-segment device is clearly limited in the number of additional digits it can display. The six alphabetic characters A–F are, of course, possible and hence a seven-segment device can also be used for displaying the sixteen hexadecimal digits. This is particularly useful for displaying the contents of various memory locations or microprocessor registers with small, single-board computers, for example.

For applications requiring the display of the full range of alphabetic and numeric (alphanumeric) digits or characters, however, it is necessary to use an enhanced type of display. One example of a display used for this purpose is the sixteen-segment display. This operates using the same principles as the seven-segment display considered earlier, but the increased number of segments means that a much wider range of characters can be displayed. This is illustrated in Figure 6.16.

Because of the enhanced range of characters that can be displayed with these devices, typical devices available use six binary input bits to specify the specific character or digit to be displayed. There are therefore sufficient combinations of the input bits (64) to be able to specify the full range of alphabetic (A–Z) and numeric (0–9) characters and also some additional characters (?,.,etc).

Again, the interface devices currently available normally have the ability to display more than one character and, as an example, a device containing a four-character display is shown in Figure 6.17. The device provides the combined latching, decoding and segment driving functions and hence a character is displayed in a specific digit position simply by outputting the six-bit binary code for the character and enabling (setting to logical 1) the appropriate latch (digit selection) input line.

6.5   Serial input and output of data

The range of input and output devices so far described are typical of those currently used for many microcomputer applications. As the level of sophistication and processing capability of the microprocessor increases, however, so the range of application areas of microcomputers also widens. It is now becoming commonplace, for example, to find microprocessors being used to build quite powerful personal computer systems and, as the processing capability of microprocessors increases, the power of these systems will continue to increase.

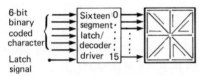

Character	Sixteen-segment representation
A	
B	
T	
Y	
Z	
0	
9	

Figure 6.16   Sixteen-segment character display

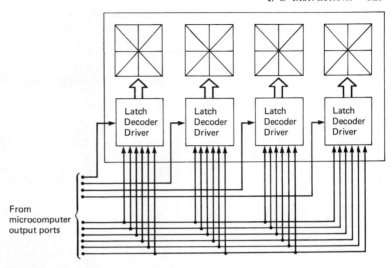

Figure 6.17   Four-character display

The most widely used device for input and output in this type of application is either a teletypewriter (teletype) or a visual display unit (VDU). Both of these devices are particularly suited to this type of application since they provide, through a full keyboard, a convenient facility for entering both strings of characters and numeric values into the microcomputer and also, through either the printed form produced by the teletypewriter or the alphanumeric display associated with the VDU, a convenient facility for recording or observing results output by the microcomputer.

As an example, a photograph of a typical commercially available VDU is shown in Figure 6.18. As can be seen, it is comprised of a keyboard (used to enter the full range of alphabetic and numeric characters and also a range of control characters) and a television type of display screen (used both to display the characters entered at the keyboard and also to display the results produced by the microcomputer). Normally, the control characters entered at the keyboard (used to enter the full range of alphabetic and numeric information on the display screen (new line, space, etc.) or indicate a particular command to the program currently running in the micro-computer (insert a character, end of character string, etc.).

Both these devices differ significantly from those discussed previously since both the data input from a teletype or VDU and the data output to these devices are both in a serial rather than a parallel form. This means that instead of the binary-coded value correspond-ing to a digit entered at the keyboard being presented to the micro-computer as a collection of binary bits on a *number* of lines, it is presented in a *time serial* manner on a single line. Similarly, the data

Figure 6.18 The VT100 visual display unit (VDU) (*Courtesy: Digital Equipment Corporation*)

to be displayed by these devices must also be output by the microcomputer in a time serial form on a single line.

A binary-coded value can be output by a microcomputer in a time serial manner by using a single line of an output port and using the microprocessor program to output the value a single bit at a time with a fixed time delay between the bits. Then, if this time period (the bit rate) is known by the receiving device, the data can be read and interpreted in a synchronised way. This is shown diagrammatically in Figure 6.19(a).

Similarly, a binary-coded value presented to a microcomputer in a time serial manner on a single input port line can be interpreted and reassembled by the microcomputer reading the state of the input line in time synchronism with the data being presented. Again, the microprocessor program must compute the (known) time delay between the presentation of each new data bit on the input line. This is shown **Q6.7** diagrammatically in Figure 6.19(b).

Character format

The number of bits used to specify a character entered at a keyboard associated with a teletype or VDU is normally seven and, indeed, internationally agreed standards have now been established which

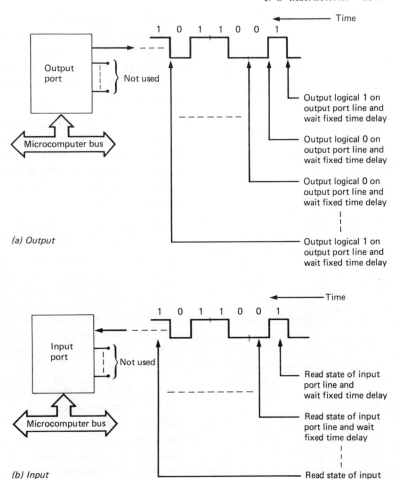

Figure 6.19 Serial I/O under program control

define the format of this data. This is given in Appendix II for reference and is known as the ISO (International Standards Organisation) 7-bit data interchange code or, alternatively, as the ASCII (American Standard Code for Information Interchange) code.

It is quite common to have a teletype or a VDU terminal connected to a microcomputer in a different physical location from the microcomputer itself. Indeed, the use of a *serial communications* approach as just outlined has been adopted specifically with this possibility in mind. Thus the terminal may be situated in one part of a laboratory and the microcomputer in another or, with the help of additional conditioning and driving circuits, in different buildings.

This physical separation of the two devices means that the transfer of data between them is prone to the introduction (through electrical interference) of so called *transmission errors*. That is, the possibility of individual bits in a character being corrupted and hence changed from logical 1 to logical 0 and *vice versa*. Clearly, the effect of this would be for the receiving device to interpret the received data in an incorrect way.

To allow for this possibility, therefore, it is normal to add an additional (eighth) bit to each 7-bit character transmitted to allow the receiving device (terminal or microcomputer) to detect whether any errors have been introduced during the transmission of the character. The eighth bit is known as a *parity bit* and it is generated and interpreted as follows.

Each 7-bit character generated by the terminal (or microcomputer) has an additional eighth bit added to it prior to transmission such that the total number of ones in the resulting 8-bit envelope, i.e. including the parity bit itself, is either even (so-called *even parity*) or odd (*odd parity*). Thus if the character to be transmitted was, say, 1000001 (ASCII character A) and even parity was being used, the eighth bit would be 0 and hence the 8-bit character transmitted would be 01000001. Similarly, if the character was 0110111 (ASCII character 7), then the eighth bit would be 1 and the character transmitted 10110111.

On receipt of each character, the microcomputer (or terminal) can then determine whether a transmission error has occurred by checking that the total number of bits received is always even (assuming even parity). If it isn't, the microcomputer can then detect this and take some appropriate corrective action for the erroneous character to be retransmitted.

Asynchronous transmission

Each character transmitted by the terminal (and microcomputer in the reverse direction) is comprised of a fixed number (eight) of bits. Once the start of a character and the bit rate are known, therefore, it is readily straightforward to synchronise the receipt and reassembly of the complete character from the incoming line. Moreover, by means of the parity bit, any transmission errors can be determined.

An important assumption, however, is that the start of each character is known. This is not always the case, however, since although it can be arranged for a microcomputer to output a string of characters at a consistent and known rate, a user entering a string of characters at a keyboard, for example, will clearly be less consistent. This mode of communication is therefore known as *asynchronous transmission* since although the individual bits which comprise a character are

(a) Character format

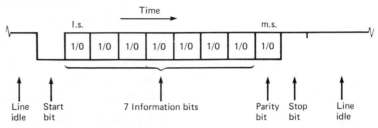

(b) Intermittent character transmission

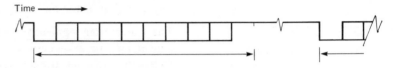

(c) Continuous character transmission

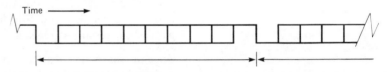

Figure 6.20 Asynchronous character format and transmission

transmitted at a fixed rate, the time between characters, and hence the rate at which characters are transmitted, is variable.

Since a transmitted character may consist of all ones or all zeros, it is necessary to have a mechanism whereby the receiving device can detect the start of each new character being transmitted. This is accomplished by always preceding the 8-bit character envelope by a single bit known as the *start bit*. Also, to ensure the transmission line always returns to the same idle state (line idle) between successive characters, each eight bit character is terminated by means of a *stop bit*. Thus the state of a typical transmission line when transmitting one or more characters is as shown in Figure 6.20.

It can be concluded from the previous discussion that the amount of actual data transmission on a line is less than the number of bit transmissions which take place. For example, assuming the above character format, for each 7-bit character it is necessary to transmit a minimum of 10 bits. The number of bit transmissions (including start and stop bits) per second is known as the *baud rate* which is always higher than the true data rate obtained from an asynchronous line. Typical standard baud rates in operation are 1,200, 2,400, 4,800 and 9,600 with visual display units and 110 baud is the standard used with a teletypewriter. It should be noted that, with the latter, it is more usual to use two stop bits rather than one and hence 110 baud is equivalent to 10 actual information characters per second.

The UART

Although it is possible to input and output serial data under program control, a more common approach is to utilise a special interface circuit known as a *UART* (universal asynchronous receiver transmitter) to perform the conversion process between the parallel binary format used at the microcomputer I/O ports and the bit serial form used on the transmission line.

A schematic diagram of a UART is shown in Figure 6.21. It is comprised of two components: one to control the transmission of the outgoing serial data stream (the transmitter) and the other to control the reception of the incoming serial data stream (the receiver).

The transmitter segment essentially performs the parallel-to-serial conversion operation between the parallel binary form output by the microcomputer and the bit serial form which is transmitted on the communication line. It is comprised, therefore, of a register whose contents are loaded in parallel and are then output in a bit serial form. It is known as the *transmitter buffer* register and is also known as a parallel-in, serial-out register.

Once a character has been loaded into the transmitter buffer register it is automatically shifted (clocked) out in time synchronism with the transmitter clock (TxC). This must be externally generated and made equal to the required baud rate. In addition, the transmitter internally generates the required parity bit for the loaded character and automatically appends the start and stop bits during transmission. Finally, after the stop bit has been transmitted, the transmitter signals to the microcomputer that the transmitter buffer register is now empty (TBE) and the next character can be loaded. This is achieved by presenting the seven bits of the character to the parallel inputs and strobing $(0 \rightarrow 1 \rightarrow 0)$ the transmitted data strobe (TDS) input.

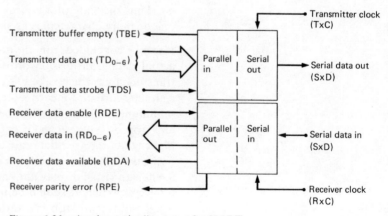

Figure 6.21   A schematic diagram of a UART

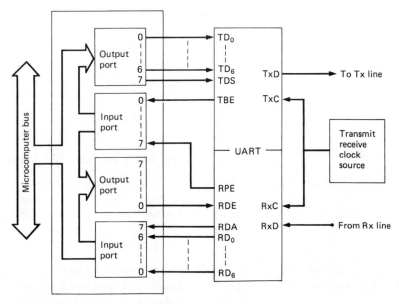

Figure 6.22 Connection of a UART via I/O ports

The receiver segment effectively performs the reverse of the transmitter segment just described. It is comprised of a serial-in, parallel-out register to perform the conversion from the serial input produced by the transmission line to the parallel output required by the microcomputer. It also contains the necessary control logic to detect the start and end of each new character. The serial data are clocked into the receiver register in bit synchronism with the receiver clock (RxC) which must therefore be made equal to the corresponding transmitter clock used within the teletype or VDU terminal.

During reception of each character, the receiver recomputes the parity of the incoming bit stream and after the stop bit has been received, determines whether or not it is the same as that contained within the received character. If the two are the same, the receiver parity error (RPE) line is reset (0) but if the two are different, the RPE line is set (1). Thus when the microcomputer reads the received character, it can readily determine if a transmission error has been detected.

In addition, when the stop bit has been received, the receiver sets the receive data available (RDA) line to logical 1 to indicate to the microcomputer that a character has been received and can now be read. The microcomputer therefore sets the receive data enable (RDE) line to logical 1 and reads the character now presented to the input port.

Because of the widespread use of UARTs in microcomputer systems, they have been designed to be connected directly to a microcomputer bus. They appear to the microprocessor, therefore, as a number of input and output ports. It is possible, however, to also connect them as an external interface circuit to a number of standard input and output ports and a typical arrangement is shown in Figure 6.22. The various control signals outlined above would then be monitored and activated under the control of the microprocessor program and the basic INput and OUTput instructions previously discussed.

6.6   Analogue input and output

As was indicated in Section 6.3, when a microcomputer is being used to control an industrial process, many of the input and subsequent output signals are digital in nature and hence, with the aid of suitable interface circuits, they may be input and output directly to or from the I/O interface unit.

In addition to digital signals, however, many industrial processes also produce signals which are analogue in nature and also some of the subsequent controlling outputs are also required in an analogue form. For example, many transducers (temperature, pressure, etc.) produce a varying (analogue) voltage which is proportional to the parameter being monitored. Similarly, a number of controlling devices require an input voltage signal the magnitude of which determines the degree of controlling action performed.

With devices of this type, therefore, it is necessary to perform a conversion operation between the digital inputs and outputs which are used at the microcomputer input and output ports and the analogue signals which are used at the controlled process. There are two types of interface circuit available for this purpose: one performs the conversion from an analogue signal to a digital signal, the other the conversion from a digital signal to an analogue signal. Both will be considered.

Digital-to-analogue conversion

The interface circuit used to convert a digital value output by a micro-computer port into its equivalent analogue form is known as a digital-to-analogue converter or DAC. Essentially, the digital inputs to a DAC are used to select a specific number of discrete voltages which are arranged in a binary weighted form. The selected voltages are then summed together to form the equivalent analogue voltage. Thus with an 8-bit converter the individual voltages would have the weighting shown in Figure 6.23 which also shows the resulting analogue output voltage for different digital inputs.

The digital inputs to the DAC can readily be obtained from an output port of the microcomputer. The latter can therefore initiate the

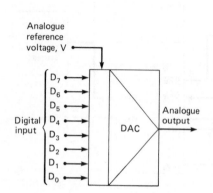

Input bit number	Analogue weighting
D_0	$V/256$
D_1	$V/128$
D_2	$V/64$
D_3	$V/32$
D_4	$V/16$
D_5	$V/8$
D_6	$V/4$
D_7	$V/2$

Digital input								Analogue output voltage
D_7	D_6	D_5	D_4	D_3	D_2	D_1	D_0	
0	0	0	0	0	0	0	0	0
0	0	0	0	0	0	0	1	$V/256$
0	0	0	0	0	0	1	0	$V/128$
0	0	0	0	0	0	1	1	$3V/256$
0	0	0	0	0	1	0	0	$V/64$
0	0	0	0	0	1	0	1	$5V/256$
1	1	1	1	1	1	0	0	$63V/64$
1	1	1	1	1	1	0	1	$253V/256$
1	1	1	1	1	1	1	0	$127V/128$
1	1	1	1	1	1	1	1	$255V/256$

Figure 6.23   Digital-to-analogue converter principles

output of an analogue voltage of a specified magnitude simply by outputting the digital equivalent of the required voltage at an output port.

Program example 6.10:   *digital-to-analogue conversion*

The program example of Figure 6.24 illustrates the effect of outputting an incrementing digital value to a DAC. It is assumed that the DAC is connected to output port 02 of the microcomputer. The digital value output is simply incremented by one within a loop and consequently, when it reaches FF (hex), it will simply reset itself to 00 (hex) and then repeat. The analogue

(a) Flowchart

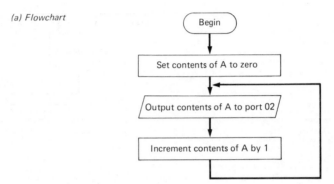

(b) Program code

Symbolic instructions	Action
LD A,00	Set contents of A to zero
REPEAT: OUT (02),A	Output contents of A to port 02
INC A	Increment contents of A by 1
JP REPEAT	Repeat

(c) Resulting output voltage

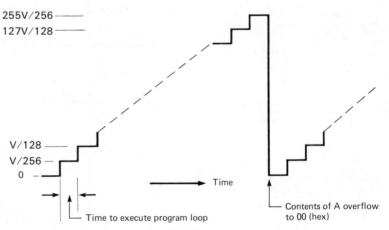

Figure 6.24   Program example 6.10

output voltage is therefore as shown earlier in Figure 6.23 and is also shown in the figure in time varying form. As each new value is output to the DAC a step increment in the resulting analogue voltage will be observed.

Q6.8

Analogue-to-digital conversion

The interface circuit used to convert an analogue voltage into its

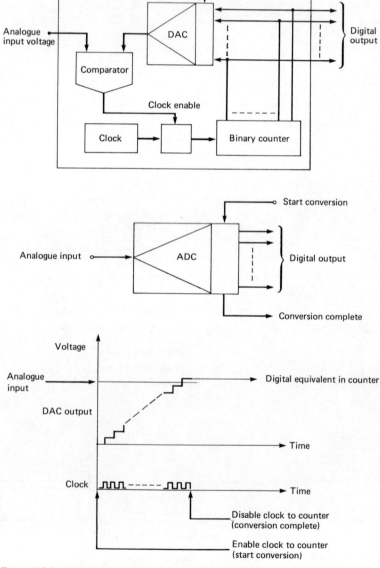

Figure 6.25   Analogue-to-digital converter principles

equivalent digital form suitable for input by a microcomputer is known as an analogue-to-digital converter or ADC.

Although a number of alternative ADC circuits are available, the basic mode of operation of these circuits is the same. An ADC does in fact incorporate a DAC within it. The circuit automatically performs an incrementing operation similar to the one illustrated in Program example 6.10. In addition, however, the circuit contains a *comparator* which, as the name implies, compares the analogue voltage generated by the DAC with the actual analogue input voltage.

If the analogue input voltage is greater than the DAC output voltage, the circuit continues incrementing the digital input to the DAC. When the DAC output voltage steps beyond the analogue input, however, the comparator detects this and stops the incrementing process. The current digital input to the DAC is then the required digital equivalent of the analogue input voltage. This is shown diagrammatically in Figure 6.25.

This mode of operation has two effects. Firstly, it is necessary to give a specific *start conversion* command to the ADC so that it can clear the digital count and start from zero for each new conversion and, secondly, the time to perform each conversion operation will vary depending on the magnitude of the analogue input voltage – the larger the voltage, the larger the number of count cycles required to increment the digital value up to the required level. The latter is important, for example, when an analogue voltage is continuously being monitored and hence converted.

To allow for both these effects, therefore, an ADC has two control lines: the start conversion input, which initiates a new conversion operation and the *conversion complete* output, which is a signal generated by the ADC to indicate when the conversion has been accomplished.

Q6.9

Questions

6.1 Assuming the connection of a set of eight switches and eight LEDs to two microcomputer ports as shown in Figure 6.2, write a program
  a to continuously read the state of the eight input switches and display the complement (inverse) of the switch states on the eight LEDs;
  b to continuously read the state of the switches and terminate when the complement of the last selected state is FF (hex).

6.2 Modify Program example 6.3 so that when the count has incremented up to FF (hex) it starts counting down (decrementing) instead of resetting to 00 (hex). Also, when it has decremented to 00 (hex), it should revert to incrementing once again and so on. Sketch the resulting waveform and hence suggest a name for it.

6.3   Write a program to store the sequence of states shown in Figure 6.7 in a table in memory and to subsequently output this sequence on a set of eight LEDs with a fixed time delay between each new state.

6.4   Write a program to store the variable time delay parameters and the associated controlled device states shown in Figure 6.8 in a table in memory and to subsequently output the specified sequence on a set of LEDs with the associated time delay between each new state.

6.5   Write a program to store the conditional input states and the corresponding controlled device states shown in Figure 6.9 in a table in memory and to subsequently read the specified input conditions from a set of eight switches before outputting the corresponding output states on a set of eight LEDs.

6.6   Write a program that simulates the input of a decimal or hexadecimal digit from a keypad as shown in Figure 6.10 by using five switches and five LEDs. The fifth switch and LED should be used to illustrate the handshake procedure shown in Figure 6.10; the remaining four switches should be used to enter the binary-coded form of the digit and the four LEDs to display the entered digit.

6.7   Write a program that outputs an 8-bit value currently stored in one of the microprocessor registers in a serial form using a single output line connected to an LED. The output bit rate should be arranged to be one bit per second.

6.8   Explain, with the aid of diagrams where appropriate, the principle of operation of a digital-to-analogue converter and hence draw a diagram to illustrate the output voltage produced by the converter if the following digital values are output by a microcomputer in a fixed time sequence:

00000000
00000001
00000010
00000100
10000000

6.9   Explain with the aid of diagrams the principle of operation of an analogue-to-digital converter. Suggest an alternative to performing a simple binary counting operation to derive the digital input to the DAC which speeds up the conversion process.

Chapter 7 Microcomputer hardware

Objectives of this chapter *When you have completed studying this chapter you should be able to:*

1 *Draw a block schematic diagram of a typical 8-bit microprocessor and show on your diagram both the flow of information on the address and data buses and the control signals generated by the microprocessor during the fetch and execute phases of some typical instruction cycles.*

2 *Appreciate the loading effect of a digital device and how this determines the number of devices that may be connected to a microcomputer bus.*

3 *Describe the function and mode of operation of a unidirectional and a bidirectional tristate buffer.*

4 *Appreciate the function of an address decoder and how it is implemented.*

5 *Draw a schematic diagram of a typical read-only memory and a random access memory device and, with the aid of timing diagrams, explain their operation.*

6 *Evolve a memory map for a typical microcomputer system and determine the associated address decoding logic required to interface the different memory devices to the microcomputer bus.*

7 *Describe, with the aid of a schematic diagram, the operation of an I/O interface unit which contains a buffered input port and a buffered output port and how the unit is interfaced to a microcomputer bus.*

8 *Understand the facilities provided by a programmable input/output (PIO) device and how it is interfaced to a microcomputer bus.*

7.1 Introduction

The basic functions and mode of operation of the three fundamental units which comprise a microcomputer – the microprocessor, the memory unit and the I/O interface unit – were discussed in Chapter 2. In addition, the structure and outline operation of the microcomputer bus was also described.

This chapter considers the operation of each of these units in detail and also the actual hardware design and associated control circuits used to implement them.

7.2   The microprocessor

A block schematic diagram of a typical 8-bit microprocessor is shown in Figure 7.1. It is similar to the figure shown previously in Chapter 2 but it now contains some of the additional registers that have been introduced in the subsequent chapters. Thus the arithmetic logic unit (ALU) has a flags register associated with it and the register unit contains a 16-bit stack pointer (SP). In addition, there is an address incrementer/decrementer unit attached to the register unit which is used, for example, to increment the program counter after each successive fetch cycle and also to increment or decrement the stack pointer after each successive stack operation.

To execute each program instruction, the microprocessor operates in a two-phase or fetch–execute mode. During the fetch phase the instruction is read from the memory unit using the address currently held in the program counter and during the execute phase the microprocessor carries out the actions necessary to implement the instruction. To illustrate the signals generated by the microprocessor and the flow of information on the data and address buses some typical fetch–execute cycles for different instruction types are now considered.

The fetch phase

As was indicated in the previous chapters, the number of bytes required to specify an instruction varies for different instruction types. For example, a simple load (move) instruction which involves two internal microprocessor registers requires only a single byte, an instruction which contains an immediate data value two bytes and an instruction which contains a memory address three bytes. Thus, depending on the number of bytes in the instruction, the microprocessor requires a variable number of basic *machine states* or *clock cycles* to fetch the instruction from the memory.

Irrespective of the number of bytes in an instruction, the first byte of the instruction is always loaded into the instruction register (IR). Then, if the instruction comprises two or more bytes, the subsequent bytes are fetched from memory and are loaded into one or more of the temporary registers (TR1, TR2 and TR3 in Figure 7.1) prior to being used during the execute phase.

The signals generated by the microprocessor and the flow of information on the address and data buses to fetch the first byte of an instruction are shown in part (a) and part (b) of Figure 7.2, respectively. The program counter (PC) is first selected and its contents enabled onto the address bus. A memory read (MEMRD) signal is then generated and the contents of the addressed location (the instruction byte) are latched into the instruction register (IR)

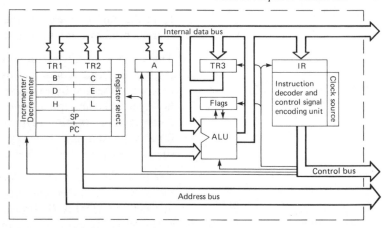

Figure 7.1 Block schematic of a typical 8-bit microprocessor

(a) Timing signals

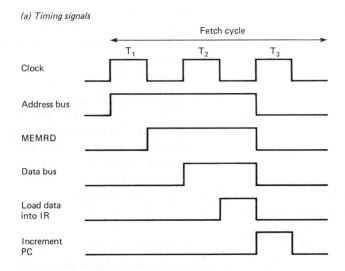

(b) Information flow

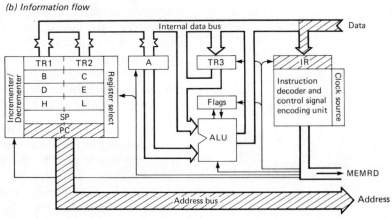

Figure 7.2 Instruction fetch – first byte

from the data bus. Finally the contents of the program counter are incremented by unity to point to the next sequential byte in memory.

The contents of the instruction register are then decoded and, if the instruction read is a single byte instruction, the execute phase is entered. If the first byte indicates it is a two or three byte instruction, however, the subsequent bytes must now be fetched from memory. As an example, the flow of information on the address and data buses for a two and a three byte instruction are shown in part (a) and part (b) of Figure 7.3, respectively.

The first example, Figure 7.3(a), assumes the second byte read from memory is an immediate data value associated with an arithmetic instruction (ADD A,7E, for example). The byte read, 7E (hex) in the example, is therefore loaded into temporary register TR3.

The second example, Figure 7.3(b), assumes the two subsequent bytes fetched from memory are two immediate data values associated with a pair of registers in the register unit (LD BC,72A8, for example). Hence the first value is loaded into temporary register TR1, 72 (hex) in the example, and the second value into temporary register TR2, A8 (hex) in the example.

After each byte of an instruction has been fetched from memory (each using a separate MEMRD signal), the contents of the program counter are incremented by unity so that the program counter always points to the next sequential byte in memory. Finally, when the required number of bytes comprising the instruction have been fetched from memory and loaded into the appropriate registers, the **Q7.1**   control unit commences the execute phase.

The execute phase

As with the fetch phase, the different instruction types require different numbers of clock cycles to execute. For example, an add immediate instruction, once the immediate data value has been loaded into temporary register TR3 during the fetch phase, requires only a single additional cycle to perform the actual addition operation whereas an add from memory instruction, say ADD A,(HL) requires two cycles: one to fetch the value from memory and the other to perform the addition.

It can be concluded from this that since the add immediate instruction involves only internal microprocessor registers, no external control signals are generated by the microprocessor. The contents of the accumulator and register TR3 are first enabled to the inputs of the ALU and the resulting sum latched into the accumulator overwriting the existing contents. The flags register is also modified according to the rules defined for this instruction.

(a) Two-byte instruction

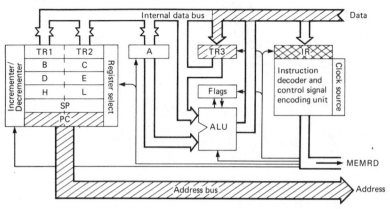

(b) Three-byte instruction

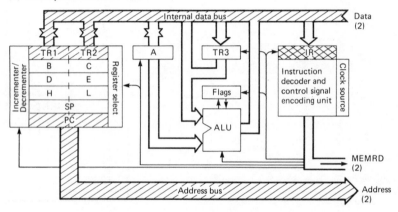

Figure 7.3 Instruction fetch – subsequent bytes

To illustrate the information flow within the microprocessor during an execute phase, the two parts of the execution cycle for the add from memory instruction are shown in parts (a) and (b) of Figure 7.4, respectively.

It can be seen from Figure 7.4(a) that during the first part of the cycle the second operand is first brought from the memory address currently specified in register pair H,L and this is temporarily stored in register TR3. Then [Figure 7.4(b)] this is added to the current contents of register A during the second part of the cycle. The sum overwrites the old contents of register A and the flags register is

Q7.2 affected according to the predefined rules.

(a) Second operand brought from address held in HL

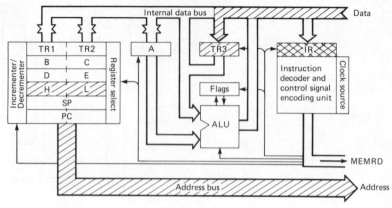

(b) This is added to current contents of A register

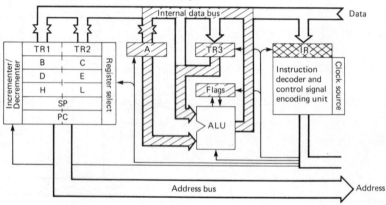

Figure 7.4 Execute phase of add from memory instruction

The start-up procedure

When power is first applied to the microcomputer, the microprocessor will begin operating immediately. The contents of the
various registers – including the program counter – will, however,
contain random information and hence, if left to do so, the microprocessor would therefore fetch the first byte from an unknown
random memory location.

For this reason, when power is first applied to a microcomputer, it is
necessary to *initialise or reset* the program counter within the microprocessor to a known state (usually zero). All microprocessors therefore utilise an external RESET control signal which when activated
performs this function. Then, if the first byte of the stored program is
arranged to be stored at this location, the microprocessor will start

executing its program immediately in a known and controlled manner. This procedure is often known as *power-on reset*.

It should be noted that the reset control signal does not normally affect the contents of any of the other microprocessor registers and hence a programmer must assume their contents are random or unknown when the stored program is first executed. When writing a program, therefore, the first actions normally performed clear or initialise any relevant microprocessor registers (the stack pointer, for example) or memory locations so that they contain known values. This is often referred to as the *initialisation sequence*.

Some of the pin assignments used with a typical 8-bit microprocessor are shown in Figure 7.5. The figure assumes the microprocessor contains a clock source (circuit) on the actual integrated circuit chip but that this requires an externally connected crystal (XTAL). This then provides a stable clock reference and, since each instruction requires a known number of clock cycles to fetch and execute, also

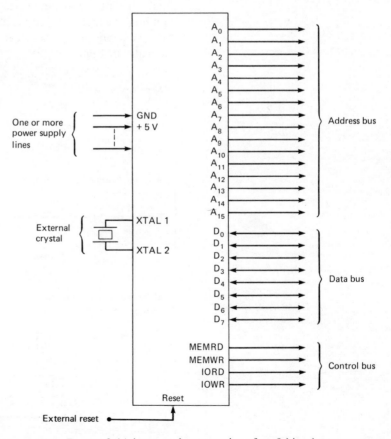

Figure 7.5 Some of the input and output pins of an 8-bit microprocessor

results in a known time duration for each instruction. This is often useful since a programmer sometimes requires to know the exact time duration for a piece of program code to be executed. Thus by counting the number of cycles required to execute each instruction in the piece of code and knowing the exact time period of each clock cycle, this can readily be determined.

Loading considerations

When using a microprocessor it is important to be aware of the electrical characteristics of the digital devices used within the microprocessor to generate the various signals – address, data and control – which appear at the output pins. These can often impose certain limitations on both the number and type of devices that may be directly connected to them.

Most digital devices used in microcomputer systems – memory devices, I/O devices, etc. – operate so that when a voltage level corresponding to a logical 0 is connected to their input, they *produce* a finite current *out* of the device (Iout). When a voltage level corresponding to a logical 1 is connected to their input, however, they *take* a finite current *into* the device (Iin). This is shown diagrammatically in Figure 7.6(a).

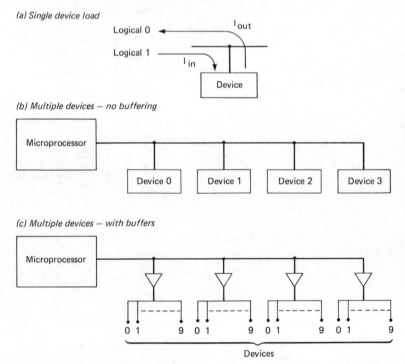

Figure 7.6   Loading considerations

Thus to function correctly with these devices, the microprocessor must be capable of receiving or *sinking* a finite current when the output signal is a logical 0 and producing or *sourcing* a finite current when the output signal is a logical 1. These currents are not normally equal, however, and usually the sink requirements of a device (logical 0) are greater than the source requirements (logical 1) – *t*ransistor *t*ransistor *l*ogic (TTL), for example.

As an example, assume the output pin of a microprocessor is able to both source and sink 1.8 mA of current and that the devices connected to this pin *produce* 0.4 mA when driven by a logical 0 and *take* 0.2 mA when driven by a logical 1. The maximum number of such devices which can be connected to the microprocessor is determined, therefore, by the logical 0 sinking capability of the microprocessor.

In this example, a maximum of four such devices could be connected to the microprocessor ($4 \times 0.4 = 1.6$ mA) and hence the latter is said to be able to *drive* four such *device loads*. Although the microprocessor would not necessarily be damaged if this number was exceeded, beyond this limit the resulting voltage level would be affected and hence the connected devices may misinterpret the logical signal being output – 0 instead of 1, 1 instead of 0. This is shown in Figure 7.6(b).

Once the drive limit of the microprocessor is reached, it then becomes necessary to add one or more additional *buffering devices* between the output pin of the microprocessor and the connected devices. This is shown in Figure 7.6(c). The figure assumes that the microprocessor can drive four devices, but the buffer circuits used can each drive up to ten similar devices.

Q7.3

Bidirectional buffers

A buffer is a device that can be used to increase the current source/sink capability of a microprocessor signal line. The buffers shown in Figure 7.6, for example, are known as *unidirectional* buffers and are therefore suitable for enhancing the drive capability of, say, the address and control lines of the microprocessor.

The data bus lines, however, in addition to being able to source or sink a particular current must also be able to operate in a *bidirectional* mode and hence special bidirectional buffers must be used when buffering a data line.

A bidirectional buffer is constructed from two tri(three)-state gates. A truth table for a *tristate gate* is shown in Figure 7.7(a). It can be seen that when the control input is logical 1, the circuit operates as a normal unidirectional buffer and the logical output is the same as the

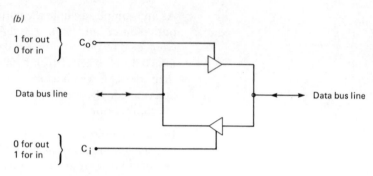

(a) Tri-state gate and truth table

In	Out	Control
1	1	1
0	0	1
1	Off	0
0	Off	0

In •―――――▷――――• Out

Control signal　C •

(b)

1 for out
0 for in } C_o o―――――

Data bus line ←――――→

0 for out
1 for in } C_i •

Data bus line ←――――→

Figure 7.7　Bidirectional buffer

input. When the control input is logical 0, however, the output is set into a *third or off state* since in this state it will neither source nor sink any current.

A bidirectional buffer can therefore be constructed using two such gates as shown in Figure 7.7(b). The figure assumes the gates are being used to buffer a data line of the microprocessor. It can be deduced from the figure that when C_o is logical 1 and C_i is logical 0, the data line will function as an output line. When C_o is logical 0 and C_i is logical 1, the data line will function as an input line. Thus C_o would typically be driven from MEMWR or IOWR and C_i from MEMRD or IORD, respectively.

Q7.4

7.3　The microcomputer bus

The microcomputer bus is used to connect all memory and I/O devices to the microprocessor. Consequently, since all these devices use the same bus lines and signals, it is necessary to ensure, firstly, that each device connected to the bus can determine when the microprocessor is wishing to communicate (transfer data) with it and, secondly, since the data bus is bidirectional, that the remainder of the devices not involved in a particular data transfer do not interfere with the transfer taking place. This can arise, for example, by a device trying to output a logical 1 on a data line whilst the microprocessor is sending a logical 0 to another device. This is shown diagrammatically in Figure 7.8.

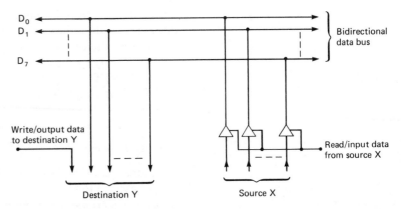

Figure 7.8 Data bus transfers

It can be deduced from the figure that to input or output a value from the data bus requires the specific source or destination of the data to be selected. This is achieved using the address bus and additional *address decoding* logic at each source and destination. The microprocessor simply outputs the identity of either the memory location or I/O device port involved in a transfer on the address bus and, by

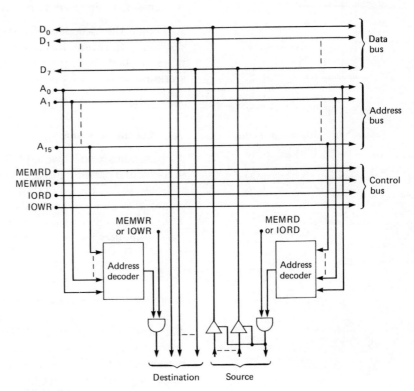

Figure 7.9 Device selection

A_1	A_0	Device selected
0	0	A
0	1	B
1	0	C
1	1	D

means of the address decoding logic, only the selected memory or I/O device responds to the request. Then, when a device determines from the state of the address bus that it is involved in a particular bus transfer, it carries out the selected action as indicated on the control lines: MEMRD or IORD for an input operation, MEMRW or IOWR for an output operation. This is shown diagrammatically in Figure 7.9.

Address decoding

An address decoder is used to select a specific memory or I/O device port connected to the microcomputer bus when it is involved in a bus transfer request. As an example, consider an address decoder which uses just two address lines, A_0 and A_1. The four possible combinations of these two lines (bits) may therefore be used as shown in Figure 7.10.

It can be seen from the figure that device A is selected (logical 1) when both A_0 and A_1 are both logical 0 and the remaining selection lines for devices B, C and D will then all be logical 0. Similarly, device B is selected when $A_0 = 1$ and $A_1 = 0$ and the other outputs will then be logical 0 and so on. As can be seen, an address decoder can readily be implemented using a combination of logical AND and inverter gates.

Although the decoders illustrated use only two inputs, larger decoders can readily be implemented using the same principles and more detailed use of these devices will be discussed in the following sections.

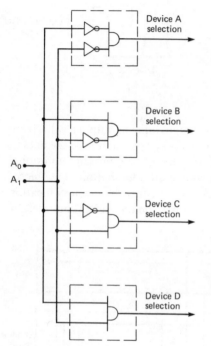

Figure 7.10   Address decoding

Q7.5

7.4   The memory unit

As was indicated in Chapter 2, the memory unit of a microcomputer is comprised of two types of device: read-only memory which is used to hold or store the list of instructions (program) and random access memory which is used to hold the data associated with this program. The internal organisation and interfacing of each type of device to the microcomputer bus is now considered.

Read-only memory

Although there are a number of different types of read-only memory available and currently in use, e.g. mask programmed (ROM) or electrically reprogrammable (EPROM), the basic operating mode of these devices is the same. The address of the required location is presented on the input address lines of the device and a memory read signal is given. The device then responds by enabling the contents of the selected location onto the data output lines. This is shown in

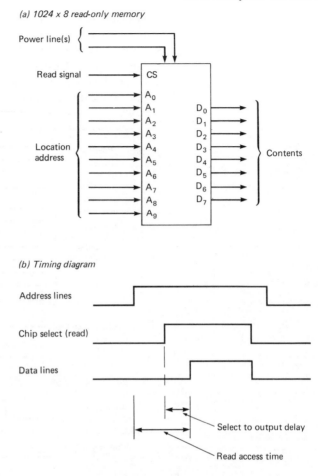

(a) 1024 x 8 read-only memory

Power line(s)

Read signal

Location
address

Contents

(b) Timing diagram

Address lines

Chip select (read)

Data lines

Select to output delay

Read access time

Figure 7.11   Read-only memory operation

Figure 7.11 which shows a typical read-only memory device which is comprised of 1,024 locations each of 8 bits.

It can be seen from the figure that the read signal is connected to the chip select (CS) input of the device since, as the name implies, this is used to indicate to the device (chip) that it has been selected. As will be seen, the read signal is not simply the MEMRD from the control bus, but rather a combination of this and the output of additional address decoding circuits.

The associated timing diagram shown in the figure shows that there is a short time delay incurred between presenting the address and chip select signals and the data being presented at the output pins. These are due to the electrical properties of the components in the actual memory array on the chip and are known as the *read access time* and the *select-to-output delay* for the device, respectively. It should be remembered that during the discussion of the timing sequence for the fetch cycle of the microprocessor (Figure 7.2), different time (clock) periods were used firstly to output the memory address and then to load the stored instruction byte into the instruction register (IR). This is specifically done to allow for the time delays illustrated which are both typically fractions of a microsecond.

Q7.6

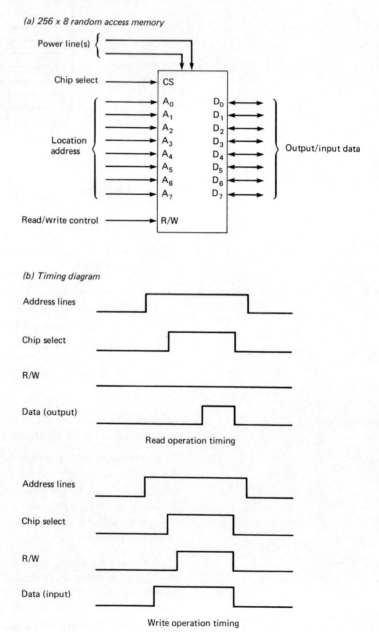

Figure 7.12 Random access memory operation

Random access memory

A schematic diagram of a typical random access memory device is shown in Figure 7.12. The device illustrated is comprised of 256 locations each of 8 bits. It can be seen from the figure that although a value may be read from or written to a location (as specified on the address input lines), the device has only a single read/write (R/W) control pin in addition to the chip select input. This is done to reduce the number of pins required for the device which in turn reduces its cost.

To perform a memory operation (read or write), the address of the appropriate location is first presented on the address input lines and the chip select input is then activated. Then, depending on the state of the R/W line, either the contents of the addressed location are read (R/W = 0) or, the current value on the data lines is written to the addressed location (R/W = 1).

To illustrate the operation of the single R/W control line, a schematic diagram of the internal organisation of a typical RAM chip is shown in Figure 7.13. The addressed location is either read from or written to as determined by the state of the R/W input. If the R/W input is low, a read operation is indicated and hence the data present on the data output lines from the memory array are enabled onto the device data lines. If the R/W input is high, however, the output from the

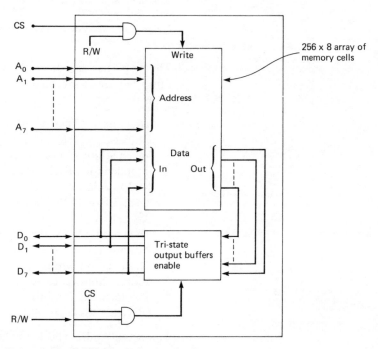

Figure 7.13 RAM internal organisation

Q7.7

memory array is disabled and instead the data present on the data lines are written into the array.

The system memory map

The memory unit of a microcomputer will typically contain several such ROM and RAM devices. For example, if an application requires, say, 3 Kbytes for the stored program, then three $1{,}024 \times 8$-bit ROMs would be required. Similarly, if the RAM section requires 1 Kbyte for data, then four 256×8-bit RAMs would be required.

When implementing the memory unit of a microcomputer, therefore, it is necessary to decide which range of memory addresses are going to be used for each device type. This is known as the *memory map* of the system and, as an example, a suitable memory map for the above system is as shown in Figure 7.14.

Q7.8

	A_{15}	A_{14}	A_{13}	A_{12}	A_{11}	A_{10}	A_9	A_8	$A_7\,A_6\,A_5\,A_4\,A_3\,A_2\,A_1\,A_0$	
0000-03FF	0	0	0	0	0	0			Chip address	ROM 1
0400-07FF	0	0	0	0	0	1			Chip address	ROM 2
0800-0BFF	0	0	0	0	1	0			Chip address	ROM 3
0C00-0CFF	0	0	0	0	1	1	0	0	Chip address	RAM 1
0D00-0DFF	0	0	0	0	1	1	0	1	Chip address	RAM 2
0E00-0EFF	0	0	0	0	1	1	1	0	Chip address	RAM 3
0F00-0FFF	0	0	0	0	1	1	1	1	Chip address	RAM 4

Figure 7.14 A system memory map

The map lists the sixteen address lines across the top and the address range required for each device type down the side. It can be deduced from the map:

1 That address lines A_{12} to A_{15} must be logical 0 for a valid memory address.
2 When address lines A_{10} and A_{11} are both logical 1, the address is intended for a RAM device otherwise the address is intended for a ROM device.
3 The specific ROM device selected is determined by address lines A_{10} and A_{11}.
4 The specific RAM device selected is determined by address lines A_8 and A_9.
5 The location within the selected ROM device is determined by address lines A_0 to A_9.
6 The location within the selected RAM device is determined by address lines A_0 to A_7.

The actual address decoding logic required to implement the above scheme is readily achieved using two 2-line to 4-line decoders similar

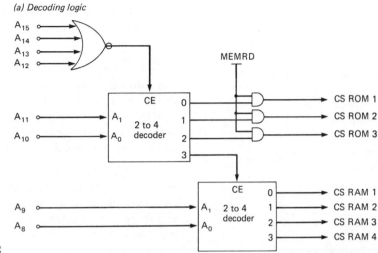

(b) Decoder truth table

CE	A_1	A_0	0	1	2	3
0	0	0	0	0	0	0
0	0	1	0	0	0	0
0	1	0	0	0	0	0
0	1	1	0	0	0	0
1	0	0	1	0	0	0
1	0	1	0	1	0	0
1	1	0	0	0	1	0
1	1	1	0	0	0	1

Figure 7.15 Memory address decoding

to those discussed earlier. A suitable arrangement is shown in Figure 7.15.

As can be seen, the address decoders have an additional chip enable (CE) input line which plays a similar role to the chip select input of a memory device. Thus when the chip enable input is logical 1, the outputs are determined by the two input combinations but when the enable is logical 0, the four outputs are all logical 0.

Hence with the first decoder, this is used in conjunction with a four-input NOR gate (OR followed by an inverter) to determine when a valid memory address is present (A_{12} to A_{15} are all logical 0) and with the second decoder, this is used to determine when a RAM device address is present.

Q7.9

7.5 The I/O interface unit

As was described in the previous chapter, the I/O interface unit is comprised essentially of a number of input and output ports. These form the interface between the microcomputer and the connected devices or interface circuits and effectively isolate the microcomputer bus from any input or output activity taking place.

The I/O interface unit is connected to the same microcomputer bus as the memory unit and hence the address bus is also used to specify the particular port involved in a data transfer. The use of separate IORD and IOWR control lines, however, allows the unit both to discriminate between memory and I/O transfers and also to determine the direction of the required transfer. Moreover, since most microcomputers provide a maximum of 256 input and/or output ports, only eight address lines are normally used to specify a port address. These are usually the least significant eight lines (A_0 to A_7) of the address bus.

Input port

The simplest form of input port is a row of eight tristate buffers as shown in Figure 7.16(a). Thus, when the microprocessor wishes to read the current value present at the input of a port, it simply outputs the address of the port on the least significant eight lines of the address bus and generates an IORD signal.

Clearly, with this approach the input device or interface circuit must ensure the data value is maintained constant at the input to the buffers until the microprocessor has read the value. An alternative and more flexible arrangement, therefore, is to incorporate an 8-bit latch into the port as shown in Figure 7.16(b).

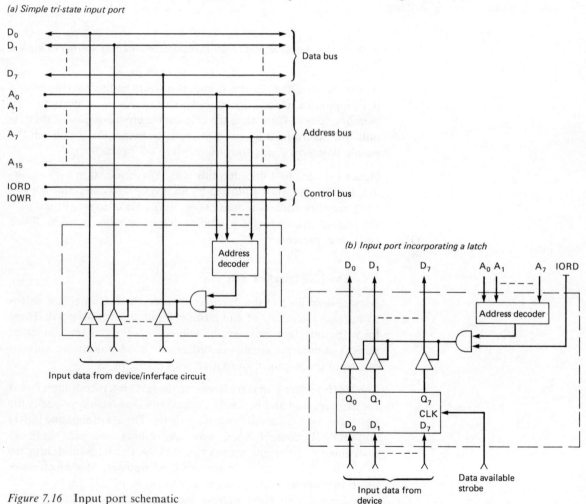

Figure 7.16 Input port schematic

A latch is a digital device used to hold or store a digital value. The value is entered or stored in the latch by presenting the value on the data inputs (D) and *strobing or clocking* the common clock input (CLK). The input value can then change but the output (Q) will remain the same until the device is clocked again.

When using an input port which incorporates a latch, therefore, the external device or circuit typically clocks a value into the latch and the microprocessor then subsequently reads the value when it is ready.

Output port

In its simplest form, an output port need only contain address decoding logic (to determine when its own address is present on the address bus) and an AND gate to combine the address selection signal with the IOWR control signal. Again, however, this would necessitate the output device firstly to be waiting for each data value to be output and secondly, being able to save or store the data value when it is output. It is usual, therefore, for an output port to contain, in addition to an address decoder, an 8-bit latch so that a data value output by the microprocessor will remain at the output port until it is subsequently accepted by the connected device or interface circuit. This is shown in Figure 7.17.

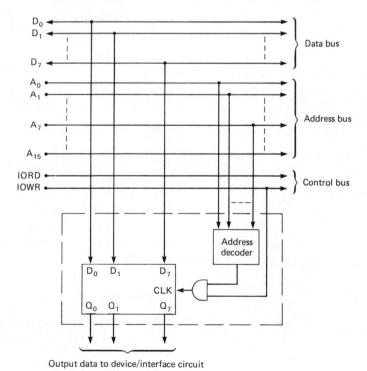

Figure 7.17 Output port schematic

Timing considerations

The timing waveforms generated by the microprocessor for an INput and an OUTput instruction are similar to those shown previously for a memory read and a memory write instruction, respectively. Thus when executing an INput instruction, the microprocessor first outputs the required port address (as specified in the instruction) on the least significant eight address lines and generates an IORD signal. The selected port then responds by gating the current value present at the tristate buffers onto the data bus and the microprocessor then loads this value into register A. A typical set of timing waveforms for an INput instruction is shown in Figure 7.18(a).

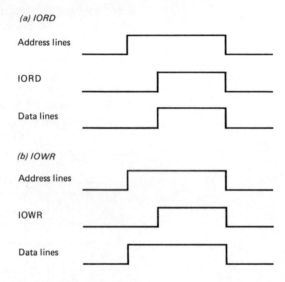

Figure 7.18 I/O timing waveforms

Figure 7.18(b) illustrates a typical set of timing waveforms for an OUTput instruction. The microprocessor first outputs the required port address (as specified in the instruction) on the address bus and the value to be output (the current contents of register A) on the data bus. It then generates an IOWR signal. The selected port then responds by clocking (latching) the value from the data bus into the port latch.

Q7.10

Programmable I/O devices

Although there are integrated circuits available to perform the basic latching and buffering functions just outlined, since the provision of I/O interfacing circuits is fundamental to most microcomputer applications, microprocessor manufacturers also provide complete I/O interface units on a single integrated circuit. Typically, these contain two or more input and/or output ports and also associated handshake control lines for each port. They are known as *programmable input/output devices or PIOs*.

The operation of the handshake procedure to coordinate the transfer of data between an I/O port and a peripheral device or interface circuit was introduced in the previous chapter. The source of the data – the microcomputer port for output and the device or interface circuit for input – asserts the DAV line when it has placed the data on the data lines and the receiver of the data indicates on the DACC line when it has accepted or read the data.

A schematic diagram of a typical PIO and its interface with the microcomputer bus is shown in Figure 7.19. The PIO illustrated is assumed

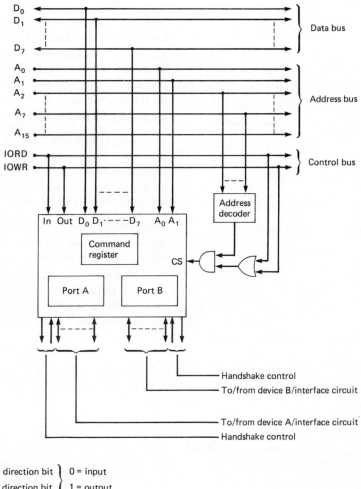

Figure 7.19 PIO schematic

to provide two bidirectional ports, A and B, together with associated handshake control lines.

The bidirectional feature of the ports means that each port can be programmed to be either an input port or an output port. The user selects the required function during the initialisation part of the program by writing either a logical 0 (for input) or a logical 1 (for output) into specific bits in the command register.

Other bits in the command register are used in conjunction with the handshake control lines. For example, assuming port A has been programmed to operate as an input port, a specific bit in the command register will be set whenever a new data value has been entered into the port latch by a connected device or interface circuit. The microprocessor, therefore, typically inputs the contents of the command register at regular intervals and examines this bit. Then, when it determines the bit has been set (and hence a new value has been entered into the port latch), it reads the value from the port. The action of reading (inputting) a value from the port by the microprocessor normally resets the appropriate bit in the command register and the PIO then generates a DACC (data accepted) strobe on the handshake control line to indicate to the connected device that a new value can be entered.

It can be concluded from the previous discussion that the microprocessor may require to both input and output to each of the port registers and also to the command register. Moreover, since the PIO contains three registers, the address decoder used must determine when any one of these addresses is present on the data bus.

As can be seen from the figure, this is normally accomplished by using the six higher-order address bits (A_2 to A_7) for the address decoder inputs and the two least significant address bits (A_0 and A_1) to indicate the specific register – Command, Port A or Port B – which is involved in the transfer. Thus, if the three addresses are 00 (hex), 01 (hex) or 02 (hex) respectively, the address decoder output will be high (logical 1) whenever the six address lines A_2 to A_7 are all logical zeros:

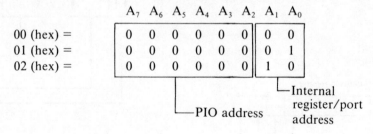

	A_7	A_6	A_5	A_4	A_3	A_2	A_1	A_0
00 (hex) =	0	0	0	0	0	0	0	0
01 (hex) =	0	0	0	0	0	0	0	1
02 (hex) =	0	0	0	0	0	0	1	0

PIO address · Internal register/port address

Q7.11

Questions

7.1 Draw a block schematic diagram of a typical 8-bit microprocessor and on your diagram show both the flow of information on the address and data buses and the control signals generated by the microprocessor to fetch the following instructions from memory:

(a) LD A,B

(b) LD B,7E

(c) LD HL,2A7E

7.2 Repeat Question 7.1 to show the execution of the following prefetched instructions:

(a) ADD A,B

(b) ADD A,48

(c) ADD A,(HL)

7.3 Explain why only a limited number of digital devices may be connected to a microprocessor and how this limit may be extended with the aid of a buffer.

7.4 Draw a truth table for a tristate gate and hence explain the operation of a unidirectional and a bidirectional buffer.

7.5 With the aid of a truth table, explain the operation of an address decoder. Show how a two-input decoder may be implemented using two-input AND gates and inverters only.

7.6 Draw a schematic diagram of a $1,024 \times 8$-bit ROM and, with the aid of a timing diagram, explain its operation. Indicate the read access time and select-to-output delay on your diagram.

7.7 Draw a schematic diagram to show the internal organisation of a typical 256×8-bit RAM and, with the aid of timing diagrams, explain its operation during a read and a write cycle.

7.8 A microcomputer system requires 4 Kbytes of ROM and 256 bytes of RAM. Determine the start and end addresses of each memory block if the two blocks occupy contiguous memory addresses starting at address 0000 (hex).

7.9 Draw a memory map for the system described in Question 7.8 and hence evolve the address decoding logic required to interface each block of memory to the microcomputer bus, assuming 1 Kbyte ROM chips and 256-byte RAM chips are to be used.

7.10 Draw a block schematic diagram of an I/O interface unit which incorporates a buffered input port and a buffered output port. Explain the operation of each block and how the unit is interfaced to the microcomputer bus.

7.11 Draw a block schematic diagram of a programmable I/O (PIO) device and explain its operation and how it is interfaced to the microcomputer bus.

Appendix I A subset of the Zilog Z80 instruction set

The following tables list some of the instructions available with the Zilog Z80 microprocessor. The instructions are presented in their symbolic notation form and the hexadecimal code of each instruction is also given. The following abbreviations are used:

A, B, C, D, E, H, L represent one of the internal microprocessor registers

(HL) represents the memory address currently held in register pair H,L

n represents an 8-bit (2 hex character) value or byte

nn represents a 16-bit (2 byte) value

The instructions selected are those which are also provided with the Intel 8080 and Intel 8085 microprocessors and hence the list is typical of the instruction set available with an 8-bit microprocessor.

1 DATA MOVEMENT INSTRUCTIONS

These instructions provide the facility for moving data from one register or memory location in the microcomputer to another. Condition flags are not affected by these instructions.

Register addressing - 8-bit data

LD	A,A 7F	B,A 47	C,A 4F	D,A 57
	A,B 78	B,B 40	C,B 48	D,B 50
	A,C 79	B,C 41	C,C 49	D,C 51
	A,D 7A	B,D 42	C,D 4A	D,D 52
	A,E 7B	B,E 43	C,E 4B	D,E 53
	A,H 7C	B,H 44	C,H 4C	D,H 54
	A,L 7D	B,L 45	C,L 4D	D,L 55

LD⎯ B,... LD⎯ C,... LD⎯ D,...

```
     ┌E,A   5F        ┌H,A   67        ┌L,A   6F
     │E,B   58        │H,B   60        │L,B   68
     │E,C   59        │H,C   61        │L,C   69
LD──┤E,D   5A    LD──┤H,D   62    LD──┤L,D   6A
     │E,E   5B        │H,E   63        │L,E   6B
     │E,H   5C        │H,H   64        │L,H   6C
     └E,L   5D        └H,L   65        └L,L   6D
```

Register addressing - 16-bit (2 byte) data

    EX DE,HL EB

Immediate addressing - 8-bit data   16-bit data

```
          ┌A,n      3E              ┌BC,nn   01
          │B,n      06              │DE,nn   11
          │C,n      0E         LD──┤HL,nn   21
          │D,n      16              └SP,nn   31
     LD──┤E,n      1E
          │H,n      26
          │L,n      2E
          └(HL),n   36
```

Direct (extended) addressing - 8-bit data   16-bit data

```
          ┌A,(nn)    3A             ┌HL,(nn)   2A
     LD──┤(nn),A    32        LD──┤(nn),HL   22
          └                         └
```

Register indirect addressing - 8-bit data

```
     ┌A,(HL)   7E          ┌(HL),A   77
     │B,(HL)   46          │(HL),B   70
     │C,(HL)   4E          │(HL),C   71
LD──┤D,(HL)   56      LD──┤(HL),D   72
     │E,(HL)   5E          │(HL),E   73
     │H,(HL)   66          │(HL),H   74
     └L,(HL)   6E          └(HL),L   75
```

2 ARITHMETIC AND LOGIC INSTRUCTIONS

Register addressing - Add*

```
        ┌ A,A  87              ┌ A,A  8F
        │ A,B  80              │ A,B  88
        │ A,C  81              │ A,C  89
ADD ─── │ A,D  82      ADC ─── │ A,D  8A
        │ A,E  83              │ A,E  8B
        │ A,H  84              │ A,H  8C
        └ A,L  85              └ A,L  8D
```

Register addressing - Subtract*

```
        ┌ A    97              ┌ A,A  9F
        │ B    90              │ A,B  98
        │ C    91              │ A,C  99
SUB ─── │ D    92      SBC ─── │ A,D  9A
        │ E    93              │ A,E  9B
        │ H    94              │ A,H  9C
        └ L    95              └ A,L  9D
```

- AND*   - OR*   - Exclusive OR*

```
        ┌ A  A7          ┌ A  B7             ┌ A  AF
        │ B  A0          │ B  B0             │ B  A8
        │ C  A1          │ C  B1             │ C  A9
AND ─── │ D  A2    OR ─── │ D  B2    XOR ─── │ D  AA
        │ E  A3          │ E  B3             │ E  AB
        │ H  A4          │ H  B4             │ H  AC
        └ L  A5          └ L  B5             └ L  AD
```

- Compare*   - Increment**   - Decrement**

```
       ┌ A  BF           ┌ A  3C              ┌ A  3D
       │ B  B8           │ B  04              │ B  05
       │ C  B9           │ C  0C              │ C  0D
CP ─── │ D  BA    INC ─── │ D  14    DEC ─── │ D  15
       │ E  BB           │ E  1C              │ E  1D
       │ H  BC           │ H  24              │ H  25
       └ L  BD           └ L  2C              └ L  2D
```

- Increment/Decrement register pair++

```
       ┌ BC  03              ┌ BC  0B
INC ─── │ DE  13      DEC ─── │ DE  1B
       └ HL  23              └ HL  2B
```

Immediate addressing - Add/Subtract*

ADD	A,n	C6		SUB	n	D6
ADC	A,n	CE		SBC	A,n	DE

- AND/OR*

AND	n	E6		OR	n	F6

- Exclusive OR/Compare*

XOR	n	EE		CP	n	FE

Register indirect addressing*

ADD	A,(HL)	86		SUB	(HL)	96
ADC	A,(HL)	8E		SBC	(HL)	9E
INC	(HL)	34		DEC	(HL)	35
AND	(HL)	A6		OR	(HL)	B6
XOR	(HL)	AE		CP	(HL)	BE

Decimal adjust accumulator*

DAA   27

Shift instructions[+]

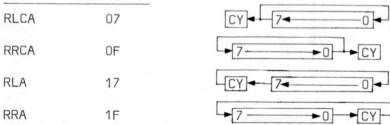

RLCA	07
RRCA	0F
RLA	17
RRA	1F

Note: *All flags affected
      **All flags except carry flag affected
      [+]Only carry flag affected
      [++]No flags affected

3 TRANSFER OF CONTROL INSTRUCTIONS

This group of instructions affects the normal sequential flow of program
execution. The following condition codes are used:

NZ = non-zero (Z = 0)	PO = parity odd (P = 0)
Z = zero (Z = 1)	PE = parity even (P = 1)
NC = non-carry (CY = 0)	P = sign positive (S = 0)
C = carry (CY = 1)	M = sign negative (S = 1)

Jump

```
JP   nn     C3          JP   PO,nn   E2
JP   NZ,nn  C2          JP   PE,nn   EA
JP   Z,nn   CA          JP   P,nn    F2
JP   NC,nn  D2          JP   M,nn    FA
JP   C,nn   DA
```

Subroutine call/return

```
CALL  nn  CD          RET  C9
```

4   INPUT/OUTPUT INSTRUCTIONS

These instructions perform an I/O operation between register A and a specified port:

```
IN  A,(n)  DB          OUT  (n),A  D3
```

5   SELECTED MACHINE CONTROL INSTRUCTIONS

Processor operations

```
NOP (no operation)  00
HALT                76
```

Stack operations

```
       ┌ AF   F5                ┌ AF   F1
       │ BC   C5                │ BC   C1
PUSH ─┤ DE   D5         POP ─┤ DE   D1
       └ HL   E5                └ HL   E1
```

Appendix II ISO/ASCII 7-bit data interchange code

The table opposite lists the complete set of codes. Groups of letters are non-printing control characters and of particular interest are CR = carriage return, LF = line feed, SP = space and DEL = delete. The parity bit shown is the even parity bit which, together with the 7-bit code, produces the complete 8 bits for each character.

Even parity bit	7-bit octal code	Char.	Even. parity bit	7-bit octal code	Char.	Even parity bit	7-bit octal code	Char.
0	000	NUL	0	053	+	0	126	V
1	001	SOH	1	054	,	1	127	W
1	002	STX	0	055	–	1	130	X
0	003	ETX	0	056	.	0	131	Y
1	004	EOT	1	057	/	0	132	Z
0	005	ENQ	0	060	∅	1	133	[
0	006	ACK	1	061	1	0	134	\
1	007	BEL	1	062	2	1	135	]
1	010	BS	0	063	3	1	136	↑
0	011	HT	1	064	4	0	137	←
0	012	LF	0	065	5	0	140	`
1	013	VT	0	066	6	1	141	a
0	014	FF	1	067	7	1	142	b
1	015	CR	1	070	8	0	143	c
1	016	SO	0	071	9	1	144	d
0	017	SI	0	072	:	0	145	e
1	020	DLE	1	073	;	0	146	f
0	021	DC1	0	074	<	1	147	g
0	022	DC2	1	075	=	1	150	h
1	023	DC3	1	076	>	0	151	i
0	024	DC4	0	077	?	0	152	j
1	025	NAK	1	100	@	1	153	k
1	026	SYN	0	101	A	0	154	l
0	027	ETB	0	102	B	1	155	m
0	030	CAN	1	103	C	1	156	n
1	031	EM	0	104	D	0	157	o
1	032	SUB	1	105	E	1	160	p
0	033	ESC	1	106	F	0	161	q
1	034	FS	0	107	G	0	162	r
0	035	GS	0	110	H	1	163	s
0	036	RS	1	111	I	0	164	t
1	037	US	1	112	J	1	165	u
1	040	SP	0	113	K	1	166	v
0	041	!	1	114	L	0	167	w
0	042	″	0	115	M	0	170	x
1	043	#	0	116	N	1	171	y
0	044	$	1	117	O	1	172	z
1	045	%	0	120	P	0	173	{
1	046	&	1	121	Q	1	174		
0	047	'	1	122	R	0	175	}
0	050	(	0	123	S	0	176	~
1	051	)	1	124	T	1	177	DEL
1	052	*	0	125	U			

Answers to Questions

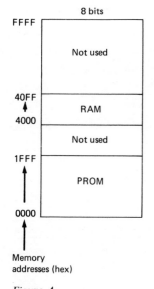

8 bits

FFFF

Not used

40FF
4000

RAM

Not used

1FFF

PROM

0000

Memory
addresses (hex)

Figure A

1.2   27 = 00011011
63 = 00111111
226 = 11100010

1.3   00101101 = 45
10101010 = 170
11011011 = 219

1.4   101 011 = 53_8
111 001 110 = 716_8
100 010 110 000 = 4260_8

1.5   C2   = 1100 0010
8E1   = 1000 1110 0001
3B2F = 0011 1011 0010 1111

2.1   Ten bits have 2^{10} or 1,024 (1K) binary combinations.
Two bits have 2^2 or 4 combinations.
Therefore, twelve bits can address 4K bytes of memory.

2.2   256 locations require 8 bits, 00 → FF (hex).
1,024 (1K) locations require 10 bits, 000 → 3FF (hex)
8,192 (8K) locations require 13 bits, 0000 → 1FFF (hex).
See also Figure A.

2.3   0000 → 07FF = 11 bits = 2K (2,048) locations for ROM.
2000 → 23FF = 10 bits = 1K (1,024) locations for RAM.

3.4   (A) = D6 (hex); (B) = 28; (C) = 00; (D) = 28; (E) = FF; (H) = 28; (L) = 00.

3.5

Symbolic instructions	Action
LD A,FF	(A) ← FF (hex)
LD B,86	(B) ← 86 (hex)
LD D,A	(D) ← FF (hex)
LD E,B	(E) ← 86 (hex)
LD HL,28A6	(HL) ← 28A6 (hex)
EX DE,HL	(DE) ← 28A6 (hex)
	(HL) ← FF86 (hex)

3.6

Symbolic instructions	Action
LD A,AA	(A) ← AA (hex)
LD (2800),A	(2800) ← AA (hex)
LD A,00	(A) ← 00 (hex)
LD HL,2800	(HL) ← 2800 (hex)
LD A,(HL)	(A) ← AA (hex)

3.7   (A) = EE (hex); (B) = EE; (C) = FF; (H) = 28; (L) = 01. Therefore

(2800) = (A) = EE
(2801) = (C) = FF

3.8

Hexadecimal code	Symbolic instructions
11 00 28	LD DE,2800
3E D6	LD A,D6
42	LD B,D
4B	LD C,E
21 FF 28	LD HL,28FF
EB	EX DE,HL
06 EE	LD B,EE
0E FF	LD C,FF
21 01 28	LD HL,2801
78	LD A,B
32 00 28	LD (2800),A
71	LD (HL),C

4.1   + 63 = 0 0111111
+112 = 0 1110000
+ 37 = 0 0100101
+111 = 0 1101111, therefore −111 = 1 0010001
+ 27 = 0 0011011, therefore − 27 = 1 1100101
+ 78 = 0 1001110, therefore − 78 = 1 0110010

4.2   0 1011011 = + 91
0 0010111 = + 23
0 1111111 = +127
1 1110111 = −(0 0001001) = − 9
1 0000101 = −(0 1111011) = −123
1 1010101 = −(0 0101011) = − 43

4.3

65 = 0 1000001	94 = 0 1011110
+24 = +0 0011000	−36 = −0 0100100
89 = 0 1011001	58 = 0 0111010

$$+28 = \ 0 \ 0011100$$
$$-28 = \ 1 \ 1100100$$
$$81 = \ 0 \ 1010001$$

$$-28 = \ 1 \ 1100100$$
$$+81 = +0 \ 1010001$$
$$\overline{+53} = \ \overline{0 \ 0110101}$$

$$+53 = \ 0 \ 0110101$$
$$-53 = \ 1 \ 1001011$$
$$23 = \ 0 \ 0010111$$

$$-53 = \ 1 \ 1001011$$
$$-23 = -0 \ 0010111$$
$$\overline{-76} = \ \overline{1 \ 0110100}$$

$$+63 = 0 \ 0111111$$
$$-63 = 1 \ 1000001$$
$$+56 = 0 \ 0111000$$
$$-56 = 1 \ 1001000$$

$$-63 = \ 1 \ 1000001$$
$$+(-56) = +1 \ 1001000$$
$$\overline{-119} = \ \overline{1 \ 0001001}$$

$$+68 = 0 \ 1000100$$
$$-68 = 1 \ 0111100$$
$$+84 = 0 \ 1010100$$
$$-84 = 1 \ 0101100$$

$$-68 = \ 1 \ 0111100$$
$$-(-84) = -1 \ 0101100$$
$$\overline{+16} = \ \overline{0 \ 0010000}$$

4.4

	0110 0111	= 67
	+0010 0100	= +24
Normal binary sum	= 1000 1011	CY = 0, AC = 0
Therefore correction	= +0000 0110	
Corrected BCD sum	= 1001 0001	= 91

	1000 1001	= 89
	=0000 1000	= +08
Normal binary sum	= 1001 0001	CY = 0, AC = 1
Therefore correction	= +0000 0110	
Corrected BCD sum	= 1001 0111	= 97

	0111 0100	= 74
	−0100 1000	= −48
Normal binary difference	= 0010 1100	CY = 0, AC = 1
Therefore correction	= +1111 1010	
Corrected BCD difference	= 0010 0110	= 26

	1001 1000	= 98
	−0011 1001	= −39
Normal binary difference	= 0101 1111	CY = 0, AC = 1
Therefore correction	= +1111 1010	
Corrected BCD difference	= 0101 1001	= 59

4.5

Symbolic instructions	*Action*
LD A,27	(A) ← 27 (hex)
ADD A,56	(A) ← 7D (hex)
DAA	(A) ← 83 (hex)
SUB 38	(A) ← 4B (hex)
DAA	(A) ← 45 (hex)

4.6

Symbolic instructions	Action
LD B,26	(B) ← 26 (hex)
LD C,69	(C) ← 69 (hex)
LD D,56	(D) ← 56 (hex)
LD A,B	(A) ← 26 (hex)
ADD A,C	(A) ← 8F (hex)
DAA	(A) ← 95 (hex)
SUB D	(A) ← 3F (hex)
DAA	(A) ← 39 (hex)

4.7

A, B	A AND B	A OR B	A XOR B
A = 1010 0111 B = 1011 1000	1010 0000	1011 1111	0001 1111
A = 1110 0010 B = 1101 1100	1100 0000	1111 1110	0011 1110
A = 0011 1100 B = 0111 0011	0011 0000	0111 1111	0100 1111
A = 0101 1010 B = 0110 1011	0100 1010	0111 1011	0011 0001

4.8

Symbolic instructions	Action
LD A,6F	(A) ← 0110 1111 CY,P not affected
LD B,A4	(B) ← 1010 0100 CY,P not affected
AND B	(A) ← 0010 0100 CY ← 0, P ← 1
OR 44	(A) ← 0110 0100 CY ← 0, P ← 0
RLCA	(A) ← 1100 1000 CY ← 0, P not affected
RLCA	(A) ← 1001 0000 CY ← 1, P not affected

5.1

Address (hex)	Hexadecimal code	Symbolic instructions		
2000	32 E7	LOOP:	LD	A,E7
2002	3C		INC	A
2003	C3 00 20		JP	LOOP
2006	–		–	

5.2 The program will loop on the instructions

```
NZERO:   DEC  B
         JP   NZ,NZERO
```

until the contents of B become zero. The program will then continue sequential execution.

5.3

Address (hex)	Hexadecimal code	Symbolic instructions		
2020	06 FF		LD	B,FF
2022	05	NZERO:	DEC	B
2023	C2 22 20		JP	NZ,NZERO
2026	–		–	

5.4 The program will loop on the instructions

```
LOOP:    INC  A
         JP   NC,LOOP
```

until the contents of A increment first to FF and then overflows to 00. The carry flag will then become set and the program will continue sequential execution. It will therefore loop 256 times.

5.5 Let B hold the running total (and the final SUM) and C the next value to be added (VAL). Flowchart is shown in Figure B and the program code is as follows:

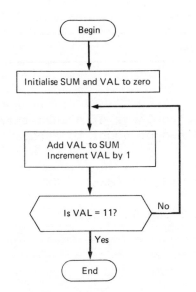

Symbolic instructions			Action
	LD	B,00	Initialise SUM and VAL to zero
	LD	C,00	
LOOP:	LD	A,B	Add VAL to SUM
	ADD	C	
	LD	B,A	
	INC	C	Increment VAL by 1
	LD	A,0B	Is VAL = 11 (dec)?
	CP	C	
	JP	NZ,LOOP	No: loop back
	HALT		Yes: end (SUM in B)

Figure B

Flowchart boxes:
- Begin
- Initialise SUM and VAL to zero
- Add VAL to SUM / Increment VAL by 1
- Is VAL = 11? → No
- Yes
- End

5.6   The flowchart is shown in Figure C and the program code is as follows:

Symbolic instructions		Action
	•	
	•	
	•	
	CP 00	Are contents of A = 0?
	JP Z,FIFTN	
	ADD A,14	No: add 20 (dec)
	JP CONT	Branch to continue
FIFTN:	ADD A,OF	Yes: add 15 (dec)
CONT:	–	Continue

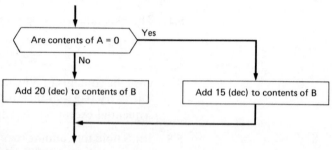

Figure C

5.7   The flowchart is shown in Figure D and the program code is as follows:

Symbolic instructions		Action
	•	
	•	
	•	
	CP 00	Are contents of A less than than or equal to zero?
	JP Z,FIFTN	
	JP M,FIFTN	
	ADD A,14	No: add 20 (dec)
	JP CONT	
FIFTN:	ADD A,OF	Yes: add 15 (dec)
CONT:	–	
	•	
	•	
	•	

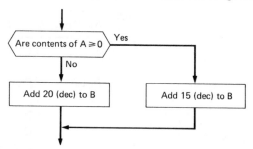

Figure D

5.8 Let the range of the numbers be held in memory address 2800 (hex) and the computed sum in memory address 2801 (hex), i.e. it is assumed the sum requires only a single byte. Let the subroutine be called SUM. The program code is as follows:

Symbolic instructions			Action
	LD	HL,2800	Initialise memory pointer to 2800 (hex)
	LD	(HL),range+1	Load number range + 1 into memory
	CALL	SUM	Call subroutine
	LD	B,(HL)	Sum of numbers in B
		.	
		.	
		.	
SUM:	LD	B,00	Initialise SUM and
	LD	C,00	VAL to zero
LOOP:	LD	A,00	Add VAL to SUM
	ADD	C	
	LD	B,A	
	INC	C	Increment VAL by 1
	LD	A,(HL)	Is VAL = number range?
	CP	C	
	JP	NZ,LOOP	No: loop back
	INC	HL	Yes: store SUM in memory block and return
	LD	(HL),B	
	RET		

5.9

Symbolic instructions			Action
	LD	C,XX	Load required delay parameter into C
DELAY:	LD	A,C	Are contents of C zero?
	CP	00	
	JP	Z,TIME	
	LD	A,64	No: loop 100 times
LOOP:	DEC	A	
	JP	NZ,LOOP	
	DEC	C	Decrement delay parameter
	JP	DELAY	and loop back
TIME:	-		Yes: time delay computed

5.10 Minimum computer delay when XX = 00 (hex)
$\qquad$ = 3 instructions or 12 μs
Maximum computer delay when XX = FF (hex)
$\qquad = 256 \times (6 + 2 \times 100) + 3$
$\qquad = 256 \times 206 = 3$
$\qquad = 52,739$ instructions
$\qquad = 210,956$ μs

5.11

Symbolic instructions			Action
	LD	D,range+1	Load number range + 1 in
	PUSH	DE	D and save on stack
	CALL	SUM	
	POP	BC	Sum of numbers in B
	•		
	•		
	•		
SUM:	POP	HL	Save return address in HL
	LD	B,00	Initialise SUM and VAL
	LD	C,00	to zero
LOOP:	LD	A,00	Add VAL to SUM
	ADD	C	
	LD	B,A	
	INC	C	Increment VAL by 1
	POP	AF	Is VAL = number range?
	CP	C	
	JP	NZ,LOOP	No: loop back
	PUSH	BC	Yes: store SUM on stack
	PUSH	HL	Restore return address on stack
	RET		Return

6.1

Symbolic instructions			Action
(a) LOOP:	IN	A,(01)	Input state of switches from port 01
	XOR	FF	Complement A
	OUT	(02),A	Output A to port 02
	JP	LOOP	Loop
(b) LOOP:	IN	A,(01)	Input state of switches
	XOR	FF	Complement A
	CP	FF	Is complemented state = FF (hex)
	JP	NZ,LOOP	No: loop
	HALT		Yes: end

6.2

Symbolic instructions			Action
REPEAT:	LD	A,00	Initialise count to 00
UP:	OUT	(02),A	Output count
	INC	A	Increment count
	CP	00	Loop if count not o/f
	JP	NZ,UP	to 00
	LD	A,FF	Initialise count to FF
DOWN:	OUT	(02),A	Output count
	DEC	A	Decrement count
	CP	FF	Loop if count not u/f
	JP	NZ,DOWN	to FF
	JP	REPEAT	Repeat

The waveform, shown in Figure E, is called a sawtooth or triangular wave.

FF

00

Figure E

6.3

Symbolic instructions			Action
	LD	HL,2800	Store table contents
	LD	(HL),15	in memory starting
	INC	HL	at address 2800
	LD	(HL),2B	(hex)
	INC	HL	

Continued

6.3 – *continued*

```
            LD      (HL),3F
            INC     HL
            LD      (HL),49
            INC     HL
            LD      (HL),54
            INC     HL
            LD      (HL),C2
            INC     HL
            LD      (HL),00
            LD      HL,2800     Initialise HL to
                                   first entry in table
REPEAT:     LD      A,(HL)      Read next state from
                                   table
            OUT     (02),A      Output next state to
                                   port 02
            CALL    DELAY       Wait delay
            INC     HL          Increment table pointer
                                   by 1
            LD      A,L         Have all states been
            CP      07             output?
            JP      NZ,REPEAT   No: repeat
            HALT                Yes: end
DELAY:      -                   Delay subroutine
            -
            .
            .
            .
            RET
```

6.4

Symbolic instructions		Action
LD	HL,2800	Table contents stored
LD	(HL),15	in consecutive memory
INC	HL	locations starting at
LD	(HL),01	address 2800 (hex)
INC	HL	
LD	(HL),2B	
INC	HL	
LD	(HL),14	
INC	HL	
LD	(HL),3F	
INC	HL	
LD	(HL),C8	
INC	HL	

Continued

```
            .
            .
            .
         LD    (HL),00
         LD    HL,2800      Initialise table pointer
REPEAT:  -                  Same as Figure 6.8
            .
            .
            .
            .
         RET
```

6.5

Symbolic instructions		Action
LD	HL,2800	Table contents stored
LD	(HL),3C	in consecutive memory
INC	HL	locations starting at
LD	(HL),2C	address 2800 (hex)
.		
.		
.		
LD	(HL),00	
LD	HL,2800	Initialise table pointer
REPEAT: -		Same as Figure 6.9
.		
.		
.		
HALT		

6.6

Symbolic instructions		Action
WTDAVS:	IN A,(01)	Input digit
	RLCA	Is DAV line set?
	JP NC,WTDAVS	No: loop
	RRCA	Yes: restore digit and DACC
	OUT (02),A	Output digit and DACC on LEDs
WTDAVR:	IN A,(01)	Input digit
	RLCA	Is DAV line reset?
	JP C,WTDAVR	No: loop
	RRCA	Yes: restore digit and DACC
	OUT (02),A	Output digit and DACC on LEDs

6.7 Let B hold the value to be output and C the current number of bits output. Assume output line is bit 0 of port 02.

Symbolic instructions			Action
	LD	B,value	Load value to be output in B
	LD	C,00	Clear bit count to 0
ONEBIT:	LD	A,B	Mask out unwanted bits
	AND	01	
	OUT	(02),A	Output bit to port 02
	CALL	ONESEC	Wait 1 sec delay
	LD	A,B	Shift next bit to l.s.
	RRCA		bit position
	LD	B,A	
	INC	C	Increment bit count
	LD	A,08	Have all bits been
	CP	C	output?
	JP	NZ,ONEBIT	No: output another bit
	HALT		Yes: end
ONESEC:	-		1 sec time delay
	.		subroutine
	.		
	.		
	RET		

7.1 See Figure F.

(a) Fetch LD A,B = 78 (hex)

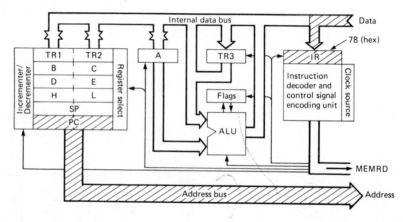

(b) Fetch LD HL,2A7E = 21 7E 2A (hex)

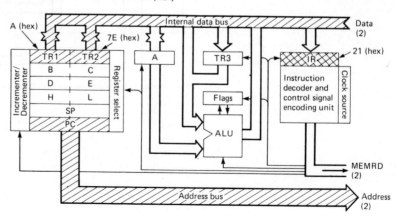

Figure F

7.2 See Figure G.

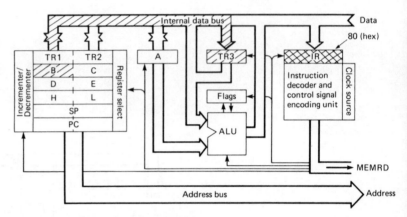

(a) Contents of B loaded into TR3

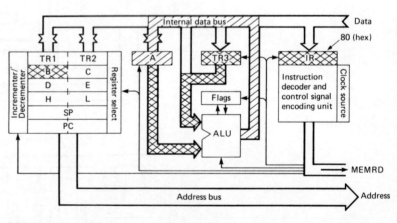

(b) Contents of A and B enabled to ALU and result loaded into A

Figure G

7.9 See Figure H. Since $1K = 10$ bits, $4K = 12$ bits and, therefore, ROM address range is 0000 to 0FFF. Also, since $256 = 8$ bits, the RAM address range is 1000 to 10FF.

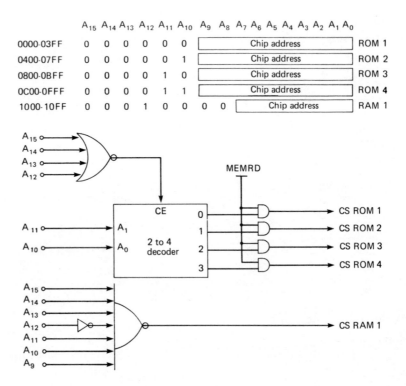

	A_{15}	A_{14}	A_{13}	A_{12}	A_{11}	A_{10}	A_9	A_8	$A_7\ A_6\ A_5\ A_4\ A_3\ A_2\ A_1\ A_0$	
0000-03FF	0	0	0	0	0	0			Chip address	ROM 1
0400-07FF	0	0	0	0	0	1			Chip address	ROM 2
0800-0BFF	0	0	0	0	1	0			Chip address	ROM 3
0C00-0FFF	0	0	0	0	1	1			Chip address	ROM 4
1000-10FF	0	0	0	1	0	0	0	0	Chip address	RAM 1

Figure H

Index